AN INTRODUCTION TO THERMODYNAMICS

AN INTRODUCTION TO THERMODYNAMICS

WITH SOME NEW DERIVATIONS BASED ON REAL IRREVERSIBLE PROCESSES

R. S. SILVER

James Watt Professor of Mechanical Engineering
University of Glasgow

CAMBRIDGE
AT THE UNIVERSITY PRESS
1971

CAMBRIDGE UNIVERSITY PRESS
Cambridge, New York, Melbourne, Madrid, Cape Town, Singapore,
São Paulo, Delhi, Dubai, Tokyo, Mexico City

Cambridge University Press
The Edinburgh Building, Cambridge CB2 8RU, UK

Published in the United States of America by Cambridge University Press, New York

www.cambridge.org
Information on this title: www.cambridge.org/9780521180139

First published 1971
First paperback edition 2010

A catalogue record for this publication is available from the British Library

Library of Congress Catalogue Card Number: 79–138380

ISBN 978-0-521-08064-4 Hardback
ISBN 978-0-521-18013-9 Paperback

To Jean

And I called it without form not that it wanted all form, but because it had such as my mind would, if presented to it, turn from, as unwonted and jarring, and human frailness would be troubled at. – and true reason did persuade me, that I must utterly uncase it of all remnants of form whatsoever, if I would conceive matter absolutely without form; and I could not; – So my mind gave over to question thereupon with my spirit, it being filled with the images of formed bodies, and changing and varying them, as it willed; and I bent myself to the bodies themselves, and looked more deeply into their changeableness, by which they cease to be what they have been, and begin to be what they were not; and this same shifting from form to form, I suspected to be through a certain formless state, not through a mere nothing; yet this I longed to know, not to suspect only.

From Confessions of St Augustine

CONTENTS

CONTENTS

CONTENTS

PREFACE

As befits the subject of thermodynamics, this book is planned as a cycle. Beginning with the concepts of work and energy we arrive at that of heat. The association with temperature is then traced and the existence of absolute temperature and of entropy is established classically via the concept of a conversion in a heat/work cycle. We then ask the question – how do we get the energy release at a temperature above ambient in order to obtain such a conversion cycle? The answer is found by using the thermodynamic concepts to predict the existence of chemical reaction, thus closing the cyclic scheme of the book. En route much of the usual introductory material of courses in thermodynamics for engineers, physicists, and chemists is generated.

The book has been largely stimulated by the observation, during a considerable industrial experience in thermodynamic design and research, of the problems faced by graduates in utilising the knowledge which they undoubtedly had gained in their university courses. I have based my teaching on methods developed from that experience. The book gives an introductory course suitable for first year or second year undergraduates, depending on level.

Although the approach will be found to be novel, it can be combined usefully with conventional practice in alternative ways according to the views of the teacher and his particular objectives. The first obvious possibility is to use it as an introduction to a course developing detailed treatment on the same general attitude. A second possibility is to give a course, going not much further than the present text, to final year students, who have already studied thermodynamics more conventionally. This has been found useful – as opening another door and as a preparation for the more flexible thinking which industrial activity requires. The third possibility, which may be preferred by some teachers, is to use it as an introductory course only, to give a general shape to the subject, and then proceed in the more usual way. It will be found that one lecture, or two at most, is sufficient to establish the connections, even although the difference in the methodology is substantial.

This brings me to comment on what the difference is, and the reasons for adopting the present approach. The first concerns the semantic and epistemological problems inevitably present in thermodynamics.

I believe these should be explicitly exposed and discussed in the development of the subject. They are actually present, and it does the student little service to be given apparently clear-cut definitions and rules without appreciation of their inter-dependence and consequent flexibility. The student brought up too rigorously on rules and definitions is in the plight of a guest of the mythical Procrustes, losing his feet or his head in order to fit the prescribed bed. One of the most plausible half-truths, or near falsehoods, is the statement 'There's no use talking unless you begin by defining what you're talking about'. In many of the most valid and important human activities it is the common process of learning that we begin with vague notions which gain precision from the discourse. In short one expects answers at the end rather than at the beginning. And so the engineering teacher will search in vain in this book for the set definitions of system, boundary, state, control-volume, etc., with which he is familiar. In the present treatment I assume only the fundamental ideas of Newtonian mechanics and kinematics as the items which are definite at the start. The rest is vague – until the discourse gives meaning to it. And that meaning will be found to illuminate the understanding of, and improve the ability to operate with, the set definitions and procedures which are indeed useful beyond the introductory level, and should preferably be given later.

A second reason or objective is to attempt to overcome the psychological disadvantage under which the mechanical engineering student particularly has been labouring for some time. For although as is well known thermodynamics began with the study of the problems of energy conversion in the heat engine, it has long since gone very much further and ranged much more widely. Unfortunately after the initial period there was a time when engineers as such remained almost entirely parochial in their attitude, limited themselves to heat engines, and derived little or no benefit nor inspiration from the grand generalisation opened up by Maxwell, Duhem, Gibbs, Boltzmann, Carathéodory, after the initial beginnings of Kelvin, Clausius and Rankine. This produced a quite justified attitude that the mechanical engineer's thermodynamics was intellectually a poor relation. The break-away from the limited heat engine teaching in mechanical engineering came only about thirty years ago, with the work of Keenan. It is this work, more than any other, which has created the current pattern of teaching in mechanical engineering thermodynamics. It was timely and crucial in bringing about the fusion of mechanical engineers with physicists, chemists, mathematicians, and chemical engineers required by the tremendous rate of post-war advance in chemical engineering, aerodynamics, and nuclear engineering. But it has not helped in re-instating mechanical engineering thermodynamics *as such* to intellectual standing. This is not to say that it is not itself intellectually advanced. It is, and the atmosphere in current engineering thermodynamics teaching which it created is also intellectually able and scrupulous. Its procedure, however, is to create *ad hoc* a rigorous formalism as described in the previous paragraph. Keenan and his colleague Hatsopoulos in their recent work are going further along these lines towards an 'axiomatic' thermodynamics. It is worthwhile, and valid, but it is not the only way.

The situation needs alternatively or additionally some approach which will show the mechanical engineer that his own professional subject range is a constituent part, of equal intellectual validity, in the whole grand structure of thermodynamics. The improvement has to come from inside engineering, not from outside. We have to see the major thermodynamic principles as organically related to and growing from necessary conceptual attitudes to the gross and palpable stuff of engineering and everyday experience.

This was of course a commonplace to the great figures of the nineteenth century – but it has been allowed to become too commonplace. Most contemporary opinion treats nineteenth century thermodynamics rather patronisingly, as for example, 'The ingenuity by which classical thermodynamics is established by way of engineering is one of the marvels in the history of science, but its significance is essentially historical' (Professor Peter Fong, *The Foundations of Thermodynamics*; New York: Oxford University Press, 1963). But in truth the significance that Fong is talking about is essentially thermodynamic, i.e. *the very fact that classical thermodynamics could be established by way of engineering tells us something important about thermodynamics*, which we miss if we write the fact off as of historical interest only. The existence and behaviour of a turbine or compressor are no less basic experimental facts than are diffusion and conduction processes. True we have to build the turbine – but the same behaviour exists in a leaf blown by the wind.

If we really understand properly that all natural phenomena are subject to the laws of thermodynamics, it is a matter of arbitrary but purposeful selection as to which kind of phenomena we begin with. And for engineering education, engineering phenomena surely constitute an appropriate start. Thus the approach in this book has an obvious parentage – or grandparentage – but as will be seen it is by no means a mere recapitulation, for it is being done consciously in a post-Gibbs and post-Keenan era. Nevertheless I do try to get back some of the baby that was thrown out with the bath-water.

The other distinguishing characteristic is that I attempt to make the development include irreversibility from the beginning. This is done by including frictional resistance to relative motion, which is a basic fact of all our mechanical experience. The treatment retains this right through so that, for example, the entropy existence theorem is proved for a real cycle including irreversibility. The inclusion of irreversibility throughout seems to me to be essential, to avoid the hiatus which the engineering student senses to exist between thermodynamic theory, as usually presented, and his actual practice. He has always a sense of unease in using state concepts, which he has been carefully taught to regard as valid only in equilibrium, in real and immediate contexts. To establish full operational confidence and flexibility, and retain adequate precision, requires a treatment showing the logical validity of quasi-equilibrium for real irreversible processes. Moreover when the student is used even from the introductory work to including irreversibility, there is much less difficulty at a later stage in making the transition to modern developments in irreversible thermodynamics.

Although the solution of arithmetic examples is a necessary part of teaching, and the whole utility of thermodynamic practice in industry requires calculation, this book is deliberately devoid of worked problems. In presenting the subject from a somewhat novel viewpoint I am concerned to emphasise its general utility as an aid to comprehension over all aspects of thermodynamics and the inclusion of a necessarily limited selection of worked examples might unnecessarily restrict attention. The student or teacher using or recommending the book whether as main or complementary reading will find no limitation, and the necessary drilling in techniques of solution is best left to the individual teacher who will prefer to devise his own worked problems to suit his particular grade of student. Moreover the real problems of industrial practice arise unformulated, and understanding contributes more to their solution than the use of standardised procedures. Nevertheless in order to give some indication of methods of use in practice, and of subsequent teaching in a full course, an Appendix is provided in which some analytic examples are given of techniques suitable for certain particular kinds of problem.

PREFACE

Although my chief concern has been to show the general sweep of the argument and hence to concentrate on the broad – almost qualitative – logic, I have also attempted to deal with some of the important issues in fairly rigorous detail – or to indicate the ways in which a rigorous treatment might be developed. Thus I have given more attention than is usual to the problems of quasi-equilibrium state and properties, since these are not only essential in engineering practice, but are also directly involved in the theoretical developments for a real irreversible process, including the entropy existence theorem. There are points in this study of equilibrium which are likely to give scope for useful further development in fundamental theory, for I have merely sketched it.

In the foregoing remarks I have said something about defects in the current usual approach to engineering thermodynamics. Nevertheless I am also fully conscious of its many virtues and strengths, and of the difficulties and risks in attempting the task of producing something different. The present book represents little more than a step on the way.

ACKNOWLEDGEMENTS

Any writer on thermodynamics owes a very great deal to the authors of previous texts even when he may adopt a different point of view. Hence in providing this short introductory book I am undoubtedly indebted to many authors of major texts which are much more detailed and comprehensive than the present work, and too numerous to mention by name.

For more direct contact I owe much to Dr Myron Tribus, who in correspondence over many years has encouraged, criticised, and commented on, my thoughts as they developed. I am also indebted particularly to my colleague, Mr J. R. Tyldesley, for many discussions and for the enthusiasm with which he has followed the ideas through and developed them in his own teaching. Dr S. K. Nisbet has also been very helpful in teaching development, and this assistance, as well as his help in preparation of the diagrams in the book, is gratefully acknowledged. To Professor H. C. Simpson of Strathclyde University, to Dr G. A. P. Wyllie of the Natural Philosophy Department and to Dr S. J. Thomson of the Chemistry Department, of this University, and to all my colleagues in this Department, who have read and commented on early drafts, my thanks are also due. The views and treatment which I have finally adopted are of course my own.

R.S.S.

Department of Mechanical Engineering,
University of Glasgow.

NOTE ON NOMENCLATURE

While there is some standardisation of symbols in common use in thermodynamics texts, there is still a good deal of variation between authors. I have used symbols which on the whole match with current practice, but the following differences should be noted.

I have adopted the general principle that properties which are primarily mechanical will be denoted by small letters. These are of course primarily, pressure and velocity – denoted by p and v respectively.

Properties which are primarily thermal are denoted by capital letters, e.g. T, U, V, H, etc. This distinction between small letters and capitals is not preserved for quantities which are not properties.

To distinguish between specific values per unit mass and the values of a total mass, I use respectively unstressed and stressed symbols, e.g. U, V, dU, dV, ΔW_t, ΔQ_t and U', V', dU', dV', $\Delta W'_t$, $\Delta Q'_t$.

These explanations and the following list should give adequate guide to the main symbols found in the text. Subsidiary nomenclature used *ad hoc* is adequately indicated in the context.

A	cross-sectional area
c	thermal capacity of unit mass
c_p	c at constant pressure
c_v	c at constant volume
c_x	c at conditions defined by suffix
F	force or Helmholtz free energy per unit mass (context establishes which is relevant)
F_f	frictional resistance force on unit mass
F_w	working force applied to unit mass
F_ϕ	potential force per unit mass
G	Gibbs free energy or potential per unit mass
H	enthalpy per unit mass
K	performance factor of refrigerator
l_T	latent thermal capacity of isothermal change per unit volume change
m, M	mass

NOTE ON NOMENCLATURE

p pressure

Q_t transmission of energy by heating, per unit mass

ΔQ_t element of Q_t

R gas constant per unit mass

R_m gas content per mole

S entropy per unit mass

s linear position co-ordinate

T thermodynamic temperature

U internal energy per unit mass

v velocity

V volume of unit mass

W_d dynamic work per unit mass, or in Chapter 6, work from chemical potential action

ΔW_d element of W_d

W_t net or total transmission of energy by work, per unit mass

ΔW_t element of W_t

x linear position co-ordinate

GREEK SYMBOLS

α coefficient of thermal expansion at constant pressure

$$\equiv \frac{1}{V}\left(\frac{\partial V}{\partial T}\right)_p$$

β coefficient of pressure variation with temperature at constant volume

$$\equiv \frac{1}{p}\left(\frac{\partial p}{\partial T}\right)_V$$

γ ratio c_p/c_v

Δ element sign for parameters which by definition are transferred quantities

ϵ total energy per unit mass

Φ potential energy per unit mass

ρ {density or / rejection ratio}
(context establishes which is relevant)

θ temperature concept in general

ψ total *transmission* of energy

1

THE MECHANICS OF THERMODYNAMICS

New knowledge does not consist so much in our having access to a new object, as in comparing it with others already known.

Colin Maclaurin

1.1 Work and energy

It is assumed that the reader is familiar with the laws of mechanics, and the concepts of work, kinetic energy, and potential energy. In this introductory section we shall simply restate for convenience some of the main points, and emphasise those aspects which are of particular importance to our subject of thermodynamics.

Work is in general defined as that which is accomplished when a force acts through a distance in its line of action. Thus if a varying force F acts on a body along the curve AB in Figure 1.1.1, and if F_s is its tangential component, and if s is the distance measured along the curve, then

$$\text{Work from } A \text{ to } B = \int_A^B F_s\, ds. \tag{1.1.1}$$

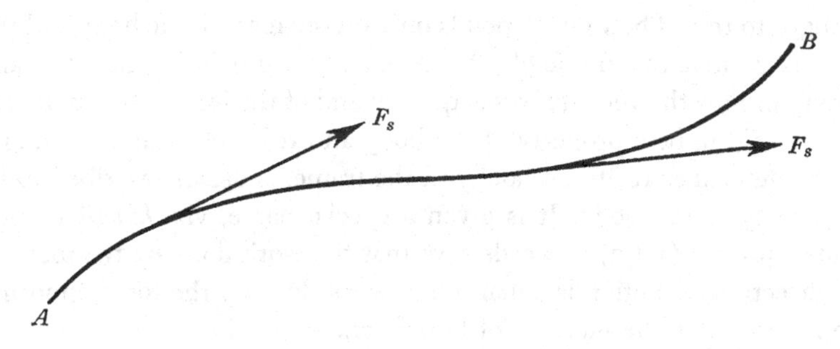

Fig. 1.1.1

This definition, which is essentially that given in most elementary textbooks, is subject to some risk of confusion. Let us suppose for example that a body of mass m is subject to two opposing forces $F_{1,s}$ and $F_{2,s}$ as shown in Figure 1.1.2.

If $F_{1,s}$ is larger than $F_{2,s}$ the body will accelerate in the direction of $F_{1,s}$ and conversely will accelerate in the direction of $F_{2,s}$ if it is the greater. What is now meant by $\int F_s \, ds$?

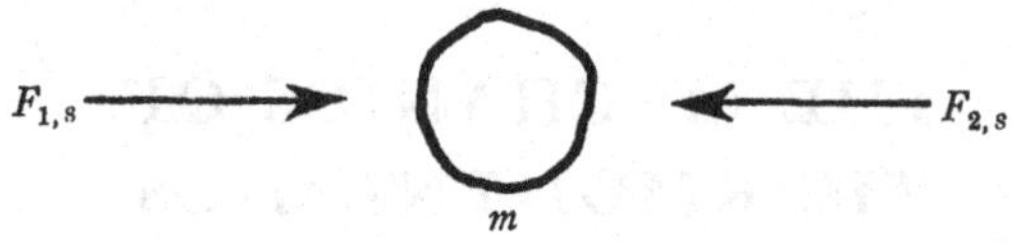

$F_{1,s}$ → ○ ← $F_{2,s}$

m

Fig. 1.1.2

From Newton's Law we know that

$$F_{1,s} - F_{2,s} = m \frac{dv}{dt};$$

$$\therefore \int_a^b F_{1,s} \, ds = \int_a^b F_{2,s} \, ds + \int_a^b m \frac{dv}{dt} \, ds$$

$$= \int_a^b F_{2,s} \, ds + \frac{mv_b^2}{2} - \frac{mv_a^2}{2}. \tag{1.1.2}$$

Thus we avoid confusion by saying that the work done *by* $F_{1,s}$ is equal to the work done *against* $F_{2,s}$ plus the increase in the value of $(mv^2)/2$. From this picture it emerges that even if $F_{1,s}$ were zero, work could still be done against $F_{2,s}$ provided the velocity diminishes, and the amount done would be given by the decrease in the value of $(mv^2)/2$. Thus a moving mass possesses a characteristic possibility of doing work, and the quantity $(mv^2)/2$ is a measure of the amount of work it might do in being brought to rest. The value depends only on the mass of the body and the velocity relative to a frame of reference, and is quite independent of any history of how the velocity was acquired, and of the forces. It may therefore be said to be a property of the body and reference frame only, and with v defined as relative velocity in the frame, is usually ascribed to be a property of the body. It is given a special name, viz. *kinetic energy*. Thus equation (1.1.2) in words says that the work done by the force in the direction of motion is equal to the work done by the force opposing the motion plus the increase of kinetic energy of the body.

In the particular case where $F_{1,s}$ is just exactly equal to $F_{2,s}$ so that there is no net force on the body, there is no acceleration and no change of kinetic energy. If motion has been started in the direction of $F_{1,s}$ at speed v it will continue and if in the direction of $F_{2,s}$ it will similarly continue. In either case we say that work is done *by* the force acting in the direction of motion *against* the force opposing the motion.

We can therefore express the general equation of motion in the following way. Let us suppose that several forces arising from different causes act simultaneously and continuously – though not necessarily in constant amount – upon a body. There will be components collinear with its direction of motion, some assisting, some opposing. We adopt the conventions indicated above that

(*a*) Forces *opposing* the direction of motion will be taken *negative*, forces assisting the direction of motion will be *positive*.

(*b*) 'Work done *by*' will refer to *positive* forces. 'Work done *against*' will refer to *negative* forces.

The equation of motion, treating all forces algebraically is

$$\sum_i F_{i,s} = m\frac{dv}{dt}.$$

$$\therefore \quad \sum_i \int_a^b F_{i,s}\,ds = \frac{mv_b^2}{2} - \frac{mv_a^2}{2}, \tag{1.1.3}$$

or leaving it in the differential form,

$$\sum_i F_{i,s}\,ds = md\left(\frac{v^2}{2}\right). \tag{1.1.4}$$

If there is no tangential acceleration $\sum_i F_{i,s}\,ds = 0$. Provided the constant value of v is not zero, work is therefore being done by some of the forces $F_{i,s}$ against the rest of them.

Some reference may be made to the other components of the forces. If the body is moving in a curved path, as shown in Figure 1.1.1, the net force on it includes a component $(mv^2)/r$, where r is the radius of curvature, directed towards the centre of curvature. Thus the other components acting on the body must combine to give this result. But since we have defined $ds = v\,dt$ to be the direction of the body motion, there is no displacement in the direction of the radius of curvature, and hence all components other than the components $F_{i,s}$ do no work.

Since each $F_{i,s}\,ds$ may be regarded as an element of work $\Delta W_{i,s}$ done by each force, we may express equation (1.1.4) alternatively as

$$md\left(\frac{v^2}{2}\right) = \sum_i \Delta W_{i,s} \tag{1.1.5}$$

or, in words, the increase of kinetic energy is the algebraic sum of the work elements done by all acting forces.

1.2 Forces and motion in a finite body

In the revision summary of Section 1.1, we have referred to the motion of a mass m, but have not indicated any finite size for it. So far as that discussion went, the mass m could have been the ideal point-particle. We now proceed to study in more detail the implications of Newtonian mechanics for a body of finite size.

For convenience in developing the principles, we begin by considering a body, such as illustrated in Figure 1.2.1, which may be thought of as a straight rod of uniform cross-sectional area A, of uniform density ρ

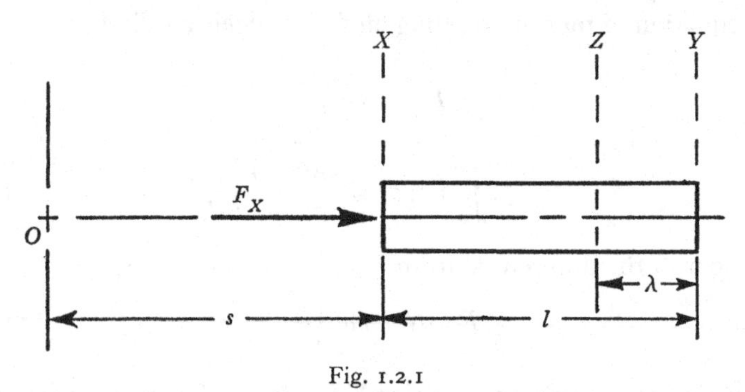

Fig. 1.2.1

and of length l between its ends X and Y. We assume that it is moving axially and first that it is subject only to the action of a single force F_X applied axially at the end X. The position of the rod in space may be indicated by the co-ordinate s, which is the distance of any defined portion of the rod such as the end X, from the origin O in the line of the axis. It is assumed for the present that the rod does not deform, and the velocity $v = ds/dt$ is uniform along the rod.

Since the total mass of the rod between the end X and the end Y is

$$m = \rho A l, \tag{1.2.1}$$

we have the equation

$$F_X = m\frac{dv}{dt} = \rho A l \frac{dv}{dt} \tag{1.2.2}$$

It is of course assumed that arrangements are such that the force F_X continues to be applied, despite the movement of the rod. Thus for example if the force F_X is applied by someone pushing the rod, he has to keep moving faster to maintain the push as the rod accelerates.

Now we may also apply the Newtonian equation to a portion of the rod of length λ which extends in front of the section plane illustrated in the figure at Z. Since the rod moves as a whole the acceleration on the portion ZY is the same as that of the whole rod or any other part of it. But since the mass is only that of length λ, the force on it is different. Denoting it be F_Z we have F_Z

$$F_Z = \rho A \lambda \left(\frac{dv}{dt}\right). \tag{1.2.3}$$

Moreover, by Newton's third law of motion we understand that since the material to the left side of the section plane Z is exerting the force F_Z forward on the ZY portion of the rod, then the material on the right side must be exerting a force F_Z *backward* on the XZ portion of the rod. Hence the equation of motion for the XZ portion is

$$F_X - F_Z = \rho A (l - \lambda) \frac{dv}{dt} \tag{1.2.4}$$

which of course agrees with (1.2.2) and (1.2.3).

We see that the essential result of the inclusion of the consideration that a real body must have a finite size is that the co-ordinate s is no longer sufficient for a complete statement. To say what the force is we must also now indicate to which portion of the body we are referring, and this requires some other co-ordinate such as l and λ and $l - \lambda$ in the examples above.

The definition of load

The situation across a section plane, such as that illustrated, where we have equal and opposite forces acting, is a characteristic one which arises throughout mechanics and a reference name is required for it. We adopt the name *load*. In the particular case shown the load is a *compressive* one. But we might obviously have the same motion in the rod maintained by a force $f_Y = F_X$ pulling in the same direction at the other end. We should then have a *tensile* load and the diagram would be as in Figure 1.2.2.

The difference between the concepts of force and of load is important and is worth further discussion to ensure a clear understanding. We note that on the section plane itself, conceived as having zero length, there is no *force* since F_Z or f_Z in opposite directions on each side cancels. But by the definition there is a *load on the section*. The amount of load is expressed as the amount of the force *on one side* of the section. The nature of the load is compressive when the directions of the equal and opposite forces on either side of the section are each *towards* the section, and is tensile when these directions are each *away from* the section.

It may be noted that in the examples given, although the accelerations are the same for the pushing and the pulling cases, the *loads* are not only different in kind, but differently distributed. Thus in Figure 1.2.1 there is the greatest compressive load F_X at the end X, corresponding to equation (1.2.2), and zero load at the end Y, corresponding to equation (1.2.3) if λ is taken to be zero, i.e. the Z section plane is moved to the end Y.

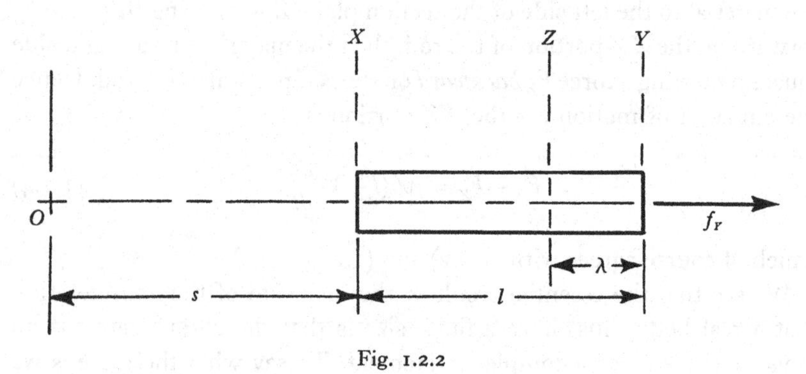

Fig. 1.2.2

But for Figure 1.2.2 the corresponding equations are

$$f_Y = \rho A l \frac{dv}{dt}, \tag{1.2.2'}$$

and

$$f_Y - f_Z = \rho A \lambda \frac{dv}{dt}. \tag{1.2.3'}$$

Equation (1.2.2') corresponds to (1.2.2), and (1.2.3') to (1.2.3). We now have the greatest tensile load $f_Y = F_X$ at the end Y when λ is zero, and zero load at the end X where λ is l, so that $f_Y - f_X = f_Y$, i.e. $f_X =$ zero.

The definition of stress

It is apparent from the foregoing that the load on a section is carried by the material of the section. We define the *stress* on the material as the *load per unit section area.*

Thus in the compression case of Figure 1.2.1 the stress at section Z is

$$\sigma_c = \frac{F_z}{A} = \rho\lambda \frac{dv}{dt}. \tag{1.2.5}$$

In the tensile case of Figure 1.2.2 the stress at section Z is similarly

$$\sigma_t = -\frac{f_z}{A}, \tag{1.2.6}$$

$$\therefore \quad \sigma_t = -\rho(l-\lambda)\frac{dv}{dt}. \tag{1.2.7}$$

Having illustrated the essential features in terms of both compressive and tensile loads and stresses, we shall from now on for simplicity omit specific discussion of the tensile case unless it happens to be particularly relevant. The discussion developed for compressive load and stress can always if required be appropriately re-interpreted for tensile conditions.

It was remarked earlier (p. 5) that the essential difference caused by considering a finite length of body in the direction of motion instead of a dimensionless particle was that we require to know to which part of the body we are referring before we can state the force situation. We now restate our problem in these terms to facilitate discussion. Figure 1.2.3

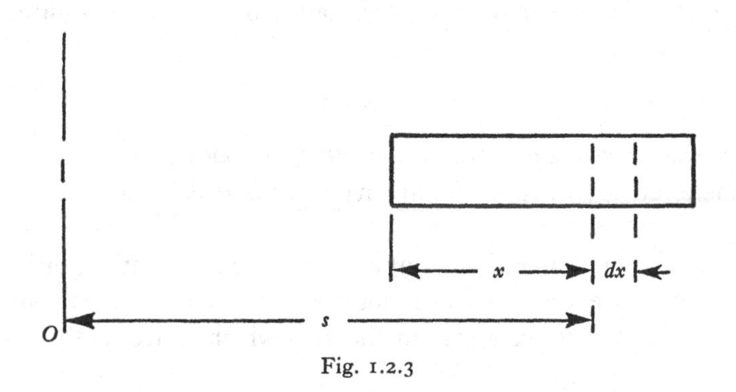

Fig. 1.2.3

shows the body, again in the form of a rod, and its position indicated by the co-ordinate s in space of any particular section of the rod. Its velocity v is ds/dt. We now define a co-ordinate x as giving the position of the same particular section with respect to the rod. We consider the equation of motion for an elementary portion of length dx of the rod.

Note that the co-ordinates s and x now refer to the same sectional plane, but are named differently in order to distinguish between an element of movement in space ds and an element of rod length dx.

The forces on the element, acting to the right, which is the assumed direction of motion, are

(1) $A\sigma$,

(2) $-A\left(\sigma+\frac{d\sigma}{dx}dx\right)$.

The sum of these is $$-A\frac{d\sigma}{dx}dx.$$

In the present case we assume that no other forces are present, and so we have, since the mass of the element is $\rho A\,dx$,

$$\rho A\,dx\left(\frac{dv}{dt}\right) = -A\frac{d\sigma}{dx}dx, \tag{1.2.8}$$

or

$$\frac{dv}{dt} = -V\frac{d\sigma}{dx}. \tag{1.2.9}$$

Thus we have the theorem that when a finite body is being accelerated under the action of normal stress forces only, there is a negative stress gradient in the direction of the acceleration.

We should emphasise again here the difference between the co-ordinates x and s, between position *in* the body and position *of* the body in space. This is perhaps brought out more clearly by pointing out that

$$\frac{dv}{dt} = \frac{d}{ds}\left(\frac{v^2}{2}\right),$$

i.e. that the stress change per *unit length of the body* (d/dx) is related to the kinetic energy change of unit mass per *unit displacement in space* (d/ds).

The theorem of equation (1.2.9) concerns the acceleration, or rate of change of kinetic energy, of a finite body, in relation to the stress distribution. We can also derive a second theorem which concerns the rate of work.

Since the load on a section at x where the stress is σ is $A\sigma$, the rate of work being done when the body is moving at velocity v is $A\sigma v$. The rate of mass flux through that section is, however, ρAv so that the work done per unit of mass carried through the section at x is σ/ρ. Thus we have the relation

σV = *work done* in conveying unit mass through a section where the stress is σ. (1.2.10)

The theorems of equations (1.2.9) and (1.2.10) are of particular interest when we consider the motion of a fluid where the stress is given by its pressure. From equation (1.2.10) we have the result that the work done in conveying a unit of mass of fluid through any section in its path of flow is pV where p is the pressure and V is the specific volume of the substance, each measured at the reference section. This quantity, being a product of the properties p and V, is itself a property of the fluid, i.e. for a given pressure and specific volume it will have the same value irrespective of the previous history of the substance. As will be seen later it is a property which is of considerable use and which enters into many calculations in practice, so it is convenient to adopt a name for it.

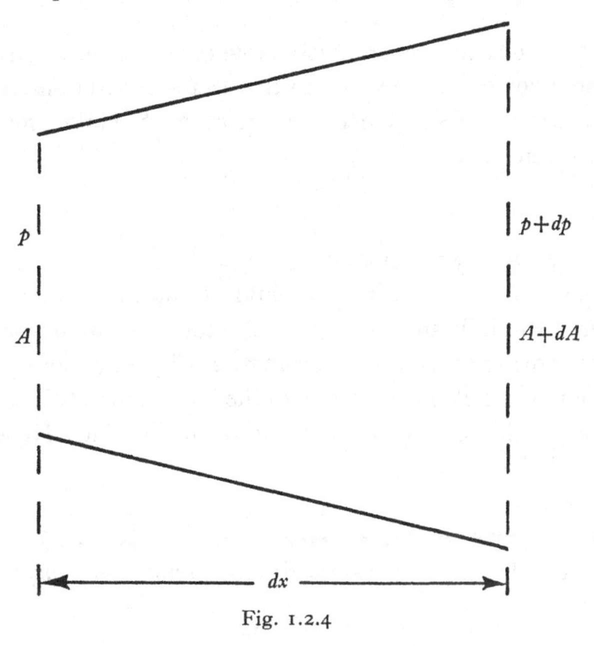

Fig. 1.2.4

The name in most frequent use at present is 'flow work'. But since it is also necessary to insist in teaching that the concept 'work' is in general *not* a property of a substance but depends on how the change process occurs, the use of the word 'work' in the combination name 'flow work' is perhaps unfortunate, and can be misleading. The term *flow function* may be used as an alternative, or we may simply say, where necessary, the pV term.

Similarly the theorem of equation (1.2.9) is also applicable to the motion of a fluid. It is, however, of interest to prove it again for this particular case because of another aspect which can be included. As shown in Figure 1.2.4, in the case of a fluid we may include a variation of area, since fluid may flow through a duct of varying cross section.

The forces on the fluid element in the direction of its motion are now

(1) Ap,

(2) $-\left(A+\dfrac{dA}{dx}\,dx\right)\left(p+\dfrac{dp}{dx}\,dx\right)$,

(3) pdA

arising from the pressure component at the walls of the duct.

The sum is

$$-A\frac{dp}{dx}\,dx$$

since second-order quantities vanish. Note therefore that the sum on the fluid element corresponds exactly to that found in Section 1.2 in establishing equation (1.2.8). Hence the equation of motion for the fluid element is given by

$$\frac{dv}{dt} = -V\frac{dp}{dx} \tag{1.2.11}$$

corresponding exactly to equation (1.2.9).

An important feature with the fluid is, however, that while the co-ordinate s still indicates merely the absolute position of the reference element of moving mass, the co-ordinate x which previously indicated only position of the element relative to the body, can now be interpreted also as the position of the element relative to the duct. Hence we can consider a relative velocity

$$v_r = \frac{dx}{dt} \tag{1.2.12}$$

for flow in the duct, as well as the absolute velocity $v = ds/dt$. We shall also have

$$u = v - v_r \tag{1.2.13}$$

where u is the absolute velocity of the duct.

Substituting from (1.2.12) in (1.2.11) we obtain therefore

$$v_r\frac{dv}{dt} = -V\frac{dp}{dt},$$

or simply

$$v_r\,dv = -V\,dp. \tag{1.2.14}$$

Using (1.2.13) gives therefore

$$v\,dv + V\,dp = -\frac{uV\,dp}{v_r}, \tag{1.2.15}$$

or

$$d\left(\frac{v^2}{2}\right) + V\,dp = -\frac{uV\,dp}{v_r}. \tag{1.2.16}$$

If u is constant, $dv = dv_r$, so that (1.2.14) gives also

$$d\left(\frac{v_r^2}{2}\right) + V\,dp = 0. \tag{1.2.17}$$

Equations (1.2.17) and (1.2.16) are of great importance to us. We note that if the duct is stationary, i.e. if $u = 0$, they become identical. In application to a stationary duct, equation (1.2.17) is known as the Bernoulli equation.

With Bernoulli's theorem for fluid flow through a stationary duct, the quantity

$$\int_1^2 V\,dp + \frac{v_2^2 - v_1^2}{2}$$

is constant. When the duct is moving, this same expression, referring to the absolute velocity of the fluid, is no longer constant, but increases if

$$\int_1^2 \frac{uV\,dp}{v_r}$$

is negative and decreases if this integral is positive.

However, equation (1.2.17) shows that the Bernoulli equation applies unmodified to the *relative* motion.

1.3 Inclusion of forces other than those arising from normal stress

The situation we have been discussing is one in which the only forces acting on the substance are those arising from normal stress. But suppose now that other forces may be present. Such a force may be either constant or be a function of the spacial position co-ordinate s. In the latter case, for any instantaneous position of the rod, this kind of force would be a function of s also. Let us consider how such possibilities may be included in our treatment.

Again we can set down the fundamental Newtonian relationship for an element of the rod of length dx and hence of mass $\rho A\,dx$ just as was done in equation (1.2.8). But if we have any forces other than those arising from normal stress we must therefore have additional terms. Since equation (1.2.8) is set up for an infinitesimal element of mass, we may assume that any such additional forces must also be represented as infinitesimals. We shall therefore denote the summation of all such force components at present conveniently as dF_s assumed to act in the direction of the motion. The equation for the element then becomes

$$-A\frac{d\sigma}{dx}\,dx + dF_s = \rho A\,dx\left(\frac{dv}{dt}\right). \tag{1.3.1}$$

From this the equation corresponding to (1.2.9) becomes

$$\frac{dv}{dt} = -V\frac{d\sigma}{dx} + \frac{dF_s}{\rho A\,dx}. \tag{1.3.2}$$

Now we define the quantity F_s by

$$F_s = \frac{dF_s}{\rho A\,dx}, \tag{1.3.3}$$

i.e. F_s is the summation *per unit mass of substance* of components in the direction of motion of the substance of any additional forces present.

$$\therefore \quad \frac{dv}{dt} = -V\frac{d\sigma}{dx} + F_s. \tag{1.3.4}$$

As an example, let us suppose that the body is vertical and is moving vertically upwards. We know then that independently of the motion, the force $F_s = -g$ acts always on unit mass. Hence equation (1.2.16) becomes

$$\frac{dv}{dt} = -V\frac{d\sigma}{dx} - g. \tag{1.3.5}$$

This corresponds to the well-known result for a stationary vertical column, where $dv/dt = 0$, so that

$$\frac{d\sigma}{dx} = -\frac{g}{V} = -\rho g, \tag{1.3.6}$$

since in this case the co-ordinate x is the height of the section. We now have a stress gradient even without acceleration.

Moreover if the body is falling freely, so that $dv/dt = -g$, we have $d\sigma/dx = 0$, i.e. we now have acceleration without a stress gradient.

The gravitational case is one example of what is called a potential force. Other potential forces, such as electrostatic, electromagnetic, etc., will also appear in F_s if they are present. A potential force may be given the particular notation F_ϕ and since displacement against such a force gives rise to potential energy Φ per unit mass, we have the relation

$$F_\phi = -\frac{d\Phi}{ds}. \tag{1.3.7}$$

Hence if potential force is the only kind of additional force present, equation (1.2.16) can be written

$$\frac{dv}{dt} = -V\frac{d\sigma}{dx} - \frac{d\Phi}{ds}. \tag{1.3.8}$$

Since $$\frac{dv}{dt} = \frac{d}{ds}\left(\frac{v^2}{2}\right),$$

this may be written $$\frac{d}{ds}\left(\frac{v^2}{2}+\Phi\right) = -V\frac{d\sigma}{dx}. \tag{1.3.9}$$

This establishes a relation between the energy increase (kinetic and potential combined) *per unit of distance moved*, and the stress gradient *per unit of distance in the substance.*

As in Section 1.2, we may apply the condition of equation (1.3.4) to the motion of a fluid. Corresponding to equation (1.2.16) we obtain

$$d\left(\frac{v^2}{2}\right) + V\,dp = F_s\,ds - \frac{uV\,dp}{v_r}. \tag{1.3.10}$$

Where F_s has a potential force only, we have

$$F_\phi = -\frac{d\Phi}{ds},$$

$$\therefore \quad d\left(\frac{v^2}{2}\right) + d\Phi + V\,dp = -\frac{uV\,dp}{v_r}. \tag{1.3.11}$$

Again, corresponding to (1.2.17), we have for the relative motion the relation

$$d\left(\frac{v_r^2}{2}\right) + \frac{v_r}{v}\,d\Phi + V\,dp = 0, \tag{1.3.12}$$

which is simply the Bernoulli equation with a potential field included.

While potential forces may be present in, or absent from, any given situation, there is another kind of force which must *always* be present in any real situation. We refer to that kind of force which we call *friction*, or *frictional force.* This is an action which occurs between two bodies in contact and in relative motion to each other, and which acts always in such a direction on each as to tend to reduce the *relative* velocity. If a rough flat stone is slid along a plank of wood lying on smooth ice, the friction between the plank and the stone will tend to slow the stone but to accelerate the plank. In slow moving fluids the frictional action appears as viscosity and in rapidly flowing fluids appears as a combination of turbulence and viscosity. The functional relationships and quantity of the frictional action, especially in fluids, requires a whole study in itself. For our purposes, however, we are not concerned with its amount, but with the certainty of its existence. Hence we must include among the possible kinds of F_s in our equation of motion the frictional force which we shall denote by $-F_f$, the negative sign being a reminder that it always acts so as to reduce the *relative* velocity.

Thus we know that the term $-F_f dx$ is always negative, i.e. $F_f dx$ is always positive.

When $-F_f$ is included among the possible forces, equation (1.3.11) becomes

$$d\left(\frac{v^2}{2}\right) + d\Phi + V\,dp = -\frac{u}{v_r}V\,dp - F_f\,ds. \tag{1.3.13}$$

We must note carefully that although $F_f dx$ is always positive, it is $F_f ds$ which appears in (1.3.13) and this may be of either sign. Suppose, for example, u is negative while v_r is positive and the arithmetical values such that $|u| > v_r$. Then v will be negative, and the frictional action of the duct on the fluid will tend to increase the absolute speed of the fluid element. The equation correctly takes account of this, since we now have ds negative, so that the term $-F_f ds$ is positive, i.e. it contributes an increment of the energy on the L.H.S. The fluid is in this case behaving as the plank in the example mentioned above, while the duct corresponds to the stone. This corresponds to the real fact that we can transmit power through a slipping clutch or a slipping belt.

Writing as before

$$\frac{ds}{dx} = 1 + \frac{u}{v_r},$$

equation (1.3.13) can be written in the form

$$d\left(\frac{v^2}{2}\right) + d\Phi + V\,dp + F_f\,dx = -\frac{u}{v_r}[V\,dp + F_f\,dx],$$

or

$$d\left(\frac{v^2}{2}\right) = -V\,dp - d\Phi - F_f\,dx - \frac{u}{v_r}[V\,dp + F_f\,dx]. \tag{1.3.14}$$

The establishment of equation (1.3.14) is the first completion point of this study. It recalls equations (1.1.4) and (1.1.5) of Section 1.1, according to which the increase of kinetic energy is the sum of the work elements of all acting forces. In equation (1.3.14) for fluid flow through a duct, we see that if u is zero the diminution of kinetic energy is simply the work done against pressure forces, potential forces, and friction. But when the duct is moving we have an additional term which in this case is

$$-\frac{u}{v_r}[V\,dp + F_f\,dx].$$

By the theorem of equations (1.1.4) and (1.1.5), this must be equivalent to the action of some force through the displacement distance ds. Since it vanishes when u is zero it is something which is brought about by the motion of the duct. It is work done by the fluid against some force opposing the motion of the duct, and which depends for its occurrence on the motion. Hence we shall define this kind of work as dynamic work W_d and so write ΔW_d for the element of dynamic work done by unit mass of fluid.

Now we are not particularly interested in the value of ΔW_d indicated in equation (1.3.14), but in the fact of the *existence* of ΔW_d when the duct is moving. This discussion has been limited to fluid flow through a straight duct moving with velocity u in the direction of its own axis. But it will equally apply to a similar duct moving in another direction, for u will then be the component of the duct velocity in the direction of fluid flow. Moreover, if the duct is curved, other components will exert an effect in the direction of the duct motion. Nevertheless there will always be the formal result that a ΔW_d will exist, proportional to the duct velocity and some function of the pressure variation in the duct. Hence equation (1.3.14) can be generalised formally as a statement for the element of dynamic work done by unit mass of any fluid element:

$$\Delta W_d = -d(\tfrac{1}{2}v^2) - d\Phi - V\,dp - F_f\,dx. \qquad (1.3.15)$$

Since $F_f\,dx$ is always positive it is convenient to refer to it as work done against friction and denote it by ΔW_f, so that

$$\Delta W_d = -d(\tfrac{1}{2}v^2) - d\Phi - V\,dp - \Delta W_f. \qquad (1.3.16)$$

1.4 The net or total work done by a fluid

We recall now the existence of the flow function pV which was defined in Section 1.2. At two successive neighbouring points in the path of an element in fluid flow, the flow function will differ by the amount $d(pV)$, i.e. the work done in conveying unit mass through a section at $s+ds$ exceeds that done in conveying unit mass through the section at s by this amount. Thus as the flow occurs, the difference work of amount $d(pV)$ is done by unit mass of fluid in the displacement. We shall define this as the element of *transport* work done by unit mass, i.e. the work done in transporting right through, conveying in at section s and conveying out at section $s+ds$. Denoting this by ΔW_{tp} we have

$$\Delta W_{tp} = d(pV). \qquad (1.4.1)$$

We are now in a position to sum up all the mechanical effects which can occur in the fluid motion. The net, or total, work done by unit mass of fluid in the elementary displacement must be the sum of the transport work and the dynamic work done by it. We shall denote this net, or total, by ΔW_t and it is given by

$$\Delta W_t = \Delta W_{tp} + \Delta W_d. \tag{1.4.2}$$

Substitution from equations (1.4.1) and (1.3.16) gives, since

$$d(pV) = p\,dV + V\,dp,$$

$$\Delta W_t = p\,dV - d(\tfrac{1}{2}v^2) - d\Phi - \Delta W_f. \tag{1.4.3}$$

For convenient comparison meantime, and for later use, we shall repeat here the equation for dynamic work ΔW_d, viz.

$$\Delta W_d = -V\,dp - d(\tfrac{1}{2}v^2) - d\Phi - \Delta W_f. \tag{1.4.4}$$

The strong similarity is apparent, with the difference only in the first terms, so that equation (1.4.2) may be satisfied.

Now it is of course obvious that if we merely began by considering a stationary element of unit mass experiencing a pressure p, the only work which it would do is $p\,dV$ if in fact its volume can change. Hence for a stationary element of fluid we could have gone straight to the equation

$$\Delta W_t = p\,dV, \tag{1.4.5}$$

which obviously achieves directly the equivalent of equation (1.4.3) without the kinetic and potential terms, which are zero if the centre of gravity does not move, and without friction.

Most introductory texts on thermodynamics do in fact go straight to this point. Dynamic work is discussed usually as a separate matter for machines and later related to net or total work. The present treatment has been adopted to demonstrate the essential unity and inter-relation, and more particularly to facilitate the thermodynamic consideration of elementary processes within machines.

Equations (1.4.3) and (1.4.4) are strictly valid only for a flow tube of infinitesimal cross section. However, assuming that p, V, v, Φ and ΔW_f refer to appropriate quasi-equilibrium values,† we can consider equa-

† The detailed mechanics of a real fluid flow is formidably complex. It requires the setting down of the Navier–Stokes equations including the Reynolds stresses for turbulence, and simple analytic solutions suitable for an introductory approach to thermodynamics are quite impracticable. But phenomenologically they are all effectively represented by a mean pressure, mean density, mean velocity, and a mean frictional resistance ascribable to an infinitesimal element of the fluid at a particular location and moving instantaneously in a particular direction. Equations (1.4.3) and (1.4.4) are then strictly valid for unit mass of substance in an infinitesimal element. To use them for practical flow considerations where the cross section of flow is finite, where local velocities, pressures, densities fluctuate, requires the assumption that a further averaging is permissible. The details of proving that such averaging is

tions (1.4.3) and (1.4.4) as relevant for unit mass of finite fluid configurations, and they can therefore be used for application to overall behaviour of fluid in machines as well as to infinitesimal flow elements.

The term $p\,dV$ is known as expansion work, and since this is the only pressure/volume term in ΔW_t, the net or total work output obtainable from a machine of *any* kind by a pressure/volume change is determined by $p\,dV$. Hence a work-producing machine can in principle be based on expansion work alone without mean fluid motion. Typical of this is the idealised piston/cylinder machine in which the charge of substance in the cylinder never changes. On the other hand, while the *net or total work* output obtainable from a pressure/volume change is *still* determined by $p\,dV$, in *any* machine through which fluid flows we have the actual occurrence of both dynamic work and transport work, even if the flow is so slow that kinetic energy terms are negligible. Thus, for example, if we merely increase the pressure on an incompressible liquid without transporting it, we do no work, since dV is zero. But if we pump it from one space into another at the increased pressure, we have to do *on* it, the dynamic work corresponding to the term $\int V\,dp$. The transport work done *by* the fluid is $\int d(pV)$, giving the net result

$$\int d(pV) - \int V\,dp = \int p\,dV = 0$$

if the fluid is incompressible. In a real liquid, which is slightly compressible, the dynamic work term $V\,dp$ done *on* the fluid is rather greater than the transport work $d(pV)$ done by the fluid, so that a small amount of net work is done *on* the fluid, corresponding to $p\,dV$ with dV negative.

Most texts on engineering thermodynamics discuss flow processes and non-flow processes as if there were a discontinuity between them. Superficially this may seem to be the case, since the dynamic work equation (1.4.4) cannot be applied when no flow is occurring. But if one keeps in mind constantly that where flow occurs both dynamic work and transport work are present one has a pattern of thought which shows continuity from flow to non-flow, i.e. from velocity v down to velocity zero, with the net or total work equation (1.4.3) always applying.

legitimate are shown in Section 2.7, Chapter 2. Meanwhile we proceed on the assumption that we can in fact usefully apply equations (1.4.3) and (1.4.4) to real finite fluid flow, with the quantities p, V, v, Φ, and the phenomenological frictional term ΔW_f interpreted as appropriate and effective averages across a normal section of mean flow. Where the conditions across such a section also vary systematically, as in the flow round a bend, a systematic averaging by integration is also required, but this introduces no problem of principle. The final effective averages for a normal section of mean flow are called quasi-equilibrium values, for reasons discussed in Section 2.7, Chapter 2.

Perhaps the most helpful way of appreciating this situation over all, is that unit mass of substance *considered in an environment moving with it* can do only the expansion work $p\,dV$ on that environment. Equation (1.4.3) shows that the net or total work is the expansion work possible against the surrounding environment considered as moving with the reference body, together with the kinetic energy and potential energy changes arising from the actual movement relative to the surroundings – less of course the frictional action. The dynamic work and the transport work are inherently concerned with motion relative to the environment, and do not exist if no such motion occurs. But expansion work against the environment is present whether or not such motion occurs.

This preponderant role of the expansion work term $p\,dV$, which appears in (1.4.3) in the principle of work output from pressure/volume changes, is the basic justification for the usual textbook approach of going straight to it. This is perhaps entirely valid for the physicist. But for the engineer who wants to see all methods of obtaining work output or requiring work input in their essential inter-relations the unity of process exhibited in equations (1.4.3) and (1.4.4) is valuable.

The above discussion of how we have to do a substantial amount of dynamic work when we are *moving* even an incompressible fluid into a region of higher pressure is one example, for a machine must be provided to do this even if the *net* work is negligible. Conversely, although we can get a substantial amount of dynamic work output from the *flow* of incompressible liquid from high pressure region to a lower, the net work output obtainable is negligible. Yet sometimes the dynamic work output can be used in particular ways. Again, this unity of process shows in the same pattern and context how work can be obtained even from an incompressible fluid at constant pressure, when both $p\,dV$ and $-V\,dp$ are zero. Then ΔW_d and ΔW_t are identical and depend on using accessible kinetic energy – such as in a windmill – or available potential energy, such as in a hydraulic turbine operating between a high level source and a low level sink of water, or a water-mill working on a mixture of kinetic and potential energy. But it also shows the converse situation which is perhaps even more important, that given the problem of trying to obtain work output from a machine with no accessible fluid at different levels, nor different pressures, and with no accessible initial velocity, the only solution is via some means by which a substance can be made to expand, i.e. by the net work $p\,dV$.

1.5 The limitations to the possibility of achieving work output

In equations (1.4.3) and (1.4.4) we have the final concise statements of how mechanical work may be obtained. If we ignore for the moment the first terms on the right-hand side of each equation we see that if we are to obtain work we must either have kinetic energy available, i.e. some initial velocity, or some initial potential energy, or both. This fact is seen in the history of mankind. The first work-producing machines depended on the presence of wind, i.e. of initial kinetic energy, or of flowing water, i.e. initial potential energy and kinetic energy. Moreover, the presence of the friction term means that the total initial sum of kinetic and potential energy must eventually be used up, unless somehow or other it is replenished.

Equations (1.4.3) and (1.4.4) may be integrated to give

$$W_t = \int_1^2 p\,dV + \frac{v_1^2 - v_2^2}{2} + \Phi_1 - \Phi_2 - \int_1^2 \Delta W_f. \qquad (1.5.1)$$

$$W_d = -\int_1^2 V\,dp + \frac{v_1^2 - v_2^2}{2} + \Phi_1 - \Phi_2 - \int_1^2 \Delta W_f. \qquad (1.5.2)$$

Thus for any process in which a fluid begins and ends with the same velocity and at the same level the possible net work output is limited to $\int p\,dV - \int \Delta W_f$, and the dynamic work output to $-\int V\,dp - \int \Delta W_f$. Now our general problem is that we must use something which is available in the ambient world and that ultimately it must settle back to the same ambient condition. We must therefore have the density and pressure also the same at beginning and end of the process, as well as velocity and level. Hence the possible net work output is limited to $\oint p\,dV - \oint \Delta W_f$, and possible dynamic work output to $-\oint V\,dp - \oint \Delta W_f$, where the symbol $\oint$ denotes an integration round a closed curve such as that shown on the pressure/volume diagram in Figure 1.5.1, where p_0, V_0 is the starting and finishing point. Now the area of this closed curve is expressible as either $\oint p\,dV$ or $-\oint V\,dp$, and hence for the closed cycle the net or total work output is equal to the dynamic work output. This corresponds to the fact that $\oint d(pV) = 0$ for the cycle, i.e. the cyclic transport work is zero since we are back where we started.

Let us denote the energy area of the closed cycle curve as shown on the diagram by the symbol ${}_p\mathbf{O}_V$. Then the work output of the cycle is

$$W_{t,c} = {}_p\mathbf{O}_V - \oint \Delta W_f. \qquad (1.5.3)$$

If we consider some ideal unreal process absolutely free from friction we should have

$$W_{t,c,0} = {}_p\mathrm{O}_V. \tag{1.5.4}$$

Thus we can say that the area ${}_p\mathrm{O}_V$ represents the ideal frictionless work output $W_{t,c,0}$ of the cycle.

Unfortunately we cannot on the same diagram illustrate $W_{t,c}$. The reason is that each point in the diagram space represents a definite pressure and specific volume condition of the substance, and while it is

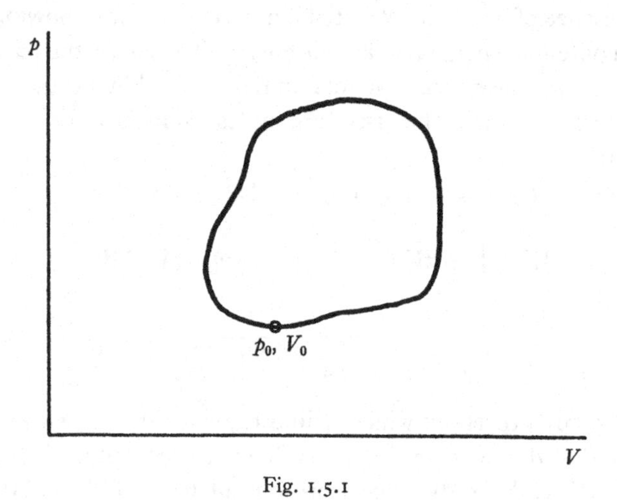

Fig. 1.5.1

tempting to draw another closed curve or less area on the diagram to show $W_{t,c}$, this would indicate different pressure and volume conditions. What can definitely be said is that

$$W_{t,c} < W_{t,c,0}, \tag{1.5.5}$$

i.e.

$$W_{t,c} < {}_p\mathrm{O}_V. \tag{1.5.6}$$

Thus the possibility of obtaining work output from a process beginning and ending with the same ambient conditions depends entirely upon whether we can vary the density of a substance in an appropriate cycle of operations in such a way that, despite the action of friction, $W_{t,c}$ is still positive.

We have now gone as far as we can with mechanics alone in describing the nature of work and the necessary conditions for obtaining it. To study whether we can, and how we can, realise the requirements set out in the previous paragraph we must turn now to other considerations.

1.6 The concepts of conservation and transmission of energy

When we examine the fundamental equation (1.4.3) for the total work done by unit mass we observe that at least a portion of it corresponds to a decrease of the summed property $\frac{1}{2}v^2 + \Phi$, i.e. to a decrease of the amount of kinetic and potential energy possessed by the mass. Now logically the fact that the amount of $\frac{1}{2}v^2 + \Phi$ has changed can be accounted for by one or other, or by each and all, of three possible concepts, viz.

(i) that some of it has been absolutely created or destroyed;

(ii) that some of it has been *transformed* to or from some equivalent other property also existing in the reference body, but not directly recognisable as either kinetic or potential energy;

(iii) that some of it has been *transmitted* to or from the reference body from or to some other distinct body possessing kinetic and/or potential energy and/or whatever other equivalent form may exist under concept (ii).

Now we are so familiar with phenomena – such as the throwing of a mass into the air and its subsequent descent – which show interplay and inter-transformation of kinetic and potential energy, that we are predisposed against the idea that they can be absolutely created or destroyed. We prefer to reject concept (i) and to adopt the contrary view that absolute creation or destruction of energy does not occur. We are then logically committed to concepts (ii) and/or (iii). But as stated in the formulation of these two concepts, transformation and transmission are not mutually exclusive, and for logical generality we must assume that both may occur.

To formalise the concept of *transformation* we shall denote by the symbol U the total value of whatever energy property or properties may exist in unit mass of a substance in some form not directly recognisable by us as kinetic and potential energy of the bulk substance, and which we shall therefore name *internal energy*. The total energy content ϵ of unit mass of a substance is then given by

$$\epsilon = U + \tfrac{1}{2}v^2 + \Phi. \tag{1.6.1}$$

The concept of *transformation* is then formally stated as saying that under conditions of constant ϵ, the three forms of energy U, $\frac{1}{2}v^2$ and Φ may be transformed to each other, i.e. when

$$\left.\begin{aligned} d\epsilon &= 0, \\ dU + d(\tfrac{1}{2}v^2) + d\Phi &= 0. \end{aligned}\right\} \tag{1.6.2}$$

The logical structure is then completed by saying that ϵ can only change as a result of *transmission*, i.e. if the total energy content of all forms contained by a reference body increases, the total energy content of everything not part of the reference body must have diminished by the same amount, and vice versa – and that some process or processes must occur to accomplish this transmission. Hence we define the quantity $\Delta\Psi_t$ as the total energy transmitted to unit mass of our reference body by whatever transmission processes there may be, and again the conservation assumption implies therefore that

$$\Delta\Psi_t = d\epsilon. \tag{1.6.3}$$

Substituting from (1.6.1) we have

$$\Delta\Psi_t = dU + d(\tfrac{1}{2}v^2) + d\Phi. \tag{1.6.4}$$

Now from the nature and definition of work we know that any statement that work is done by one entity implies that the same amount is done *on* some other entity or entities. In this sense we can therefore accept that work is a transmission process which can meet the logical requirements of concept (iii) above.

Hence we may conclude that $-\Delta W_t$, i.e. the negative of ΔW_t, which represents the net or total work done *on* unit mass of a substance, is the total amount of energy which can be transmitted to the substance by the work process. We recall however that from equation (1.4.3)

$$-\Delta W_t = -p\,dV + d(\tfrac{1}{2}v^2) + d\Phi + \Delta W_f. \tag{1.6.5}$$

Comparing (1.6.5) with (1.6.4) we see therefore that $-\Delta W_t$ is not in general equal to $\Delta\Psi_t$, i.e. that the total energy transmitted by the work process is not by itself in general able to account for the full energy transmission to the substance. The equations show that $-\Delta W_t$ is equal to $\Delta\Psi_t$ only provided the quantity $dU + p\,dV - \Delta W_f$ is zero. Hence we are forced to conclude that, if we are to assume that energy is a conserved quantity, there must exist in general some other process of energy transmission quite distinct from the work process, and that the total quantity of energy transmitted by such means in any change amounts to

$$dU + p\,dV - \Delta W_f.$$

Denoting the element of energy transmitted by this alternative process as ΔQ_t, we have therefore the relation

$$\Delta Q_t = dU + p\,dV - \Delta W_f. \qquad (1.6.6)$$

Note that this is an equation which gives the amount of the energy transmitted by the process. The presence of the work amount terms $p\,dV$ and ΔW_f must not be taken to indicate that the work process – i.e. the action of a force through a distance – is in any way involved in the process by which ΔQ_t occurs. All Newtonian force and distance effects are included in ΔW_t and hence ΔQ_t must occur by some entirely different means which is not describable in these terms.

Now of course the reader is already aware that this other process of energy transmission will be identified with that phenomenon which we call heating. He is also aware that in our lay experience from childhood we become just as familiar with the phenomena of 'heat' as with those of 'work', and that historically the principle of conservation of energy was perhaps arrived at by combining these two classes of phenomena. He is also probably aware that usually thermodynamic texts begin with 'heat' and 'work' and thence go to the principle of conservation of energy, and he may wonder why a different procedure is adopted in this book.

The basic logical situation is that there are *three* physical relations which have in fact already appeared as equations in our text, viz.

A *Conservation of energy*

$$\Delta \Psi_t = d\epsilon = dU + d(\tfrac{1}{2}v^2) + d\Phi; \qquad (1.6.3)$$

B *Mechanical transmission of energy*

$$\Delta W_t = p\,dV - d(\tfrac{1}{2}v^2) - d\Phi - \Delta W_f; \qquad (1.4.3)$$

C *Thermal transmission of energy*

$$\Delta Q_t = dU + p\,dV - \Delta W_f; \qquad (1.6.6)$$

and that only two of them are independent.

The usual procedure in introductory texts in thermodynamics is to present first separate discussions of our knowledge of thermal and of mechanical phenomena and then to proceed to conservation of energy, i.e. to begin with relations *B* and *C*. This has the satisfaction of starting from two recognised types of experience. But it carries with it the *separate* systems of everyday intuitive ideas, and the moulding of both of these simultaneously into one overall logical conceptual pattern is not easy. That this is so is confirmed by the fact that it took some 200 years after Newton before a consistent scheme of ideas of 'heat' and its relation to 'work' and energy was developed.

The difficulty is that when we begin with both B and C, which are descriptions of separate phenomena, we cannot deduce A from them. A is a postulate made additionally. The necessary steps in the argument then have to be:

(i) To assume that B and C are each transmitting the same thing – energy.

(ii) To assume that the amount of this is conserved.

(iii) To take a decision either to define the amount in terms of the phenomena of B or of the phenomena of C.

For many years the confusion of defining energy in terms of both B and C persisted but has now been abandoned.

In modern science the decision is made quite definitely to give a definition in terms of B, i.e. energy is defined in terms of force and displacement and *not* in terms of C (thermal capacity and temperature change). Through this definition, mechanics attains a priority in the conceptual structure of thermodynamics.

The approach adopted in this book can be seen from one point of view as merely recognising fully the implications of that decision.

The second reason for the present approach is that all modern teaching emphasises that 'heat' is *not*, as used to be said, a form of energy, and that neither is 'work', but that both are processes by which energy is transferred. This sense of process is not readily apparent when we begin in the usual way by introducing the phenomena of 'heat' in terms of thermometry, properties, etc. But with the procedure adopted in this book of beginning with mechanics, the first major result is that there must exist some process of energy transmission entirely different in nature from the mechanical process. The present discussion has proceeded from the 'preview' that we are going to identify this other process as thermal, and provided this identification is made, we establish right at the beginning of our studies the central core of emphasis in modern teaching of 'heat' as a process of energy transmission. As will be seen we can then go on to the discussion of thermal phenomena, thermometry, properties, etc.

The reader is therefore invited to suspend meantime his awareness that what we call 'heat' or thermal phenomena constitute the other side of our story. For to make our point fully we have now to return to equation

(1.6.6), and, in our main argument, *prove* that the energy transmission process ΔQ_t which has been deduced to exist can in fact be identified as thermal.†

† *Electric, magnetic, and capillary effects:* The beginner may wonder why no mention has been made of such matters as surface tension and magnetic and electric effects. The situation regarding these is outlined in the last chapter (p. 126). Meanwhile we may simply state that all of these come into the mechanical side of the preceding discussion, since all manifest force and transmit energy by the work mode, by the action of such force through a conjugate displacement. The result is that stresses other than the fluid pressure p must be included when electric, magnetic, and capillary effects are present. In common with most books on introductory thermodynamics we exclude these from all the following discussion until the outline in the last chapter.

2

THE LAWS OF THERMODYNAMICS

Some laws are unjust; some are just;
Others are just so.
Anon.

2.1 The identification of the other process of energy transfer

We have seen that some process of energy transfer must exist which operates by some means quite different from the work process, and that the *amount* of energy transmitted by this means is given by equation (1.6.6). We now seek to identify this process with something known in our experience. The steps by which this can be done with some definiteness are as follows.

First we note that equation (1.6.6) may be written in the form

$$\Delta Q_t + \Delta W_f = dU + p\,dV. \tag{2.1.1}$$

From this we can deduce that whatever ΔQ_t is, the results of it on the internal energy and specific volume of the mass which receives the transmitted energy are of the same nature as would be observed by doing work against friction. For, according to (2.1.1) a specified change $dU + p\,dV$ can be brought about either by large ΔQ_t and small ΔW_f, or vice versa. We cannot be more precise yet since, until we have identified the process which gives ΔQ_t, we cannot say whether or not the process is occurring in any particular circumstances. But we have our first clue since we know that if we now, keeping other conditions the same, introduce work only against friction, we shall observe the same kind of result as would be given by an increase of energy transfer by the unknown process. Now we are aware throughout our experience, again from childhood, that what we observe when we do work against friction is most frequently what we call a rise of temperature.

We do not require any precise measurement nor even precise definition of what we mean by temperature at this stage of our logic. It suffices that it is something we can detect as resulting from work done against friction, and which we can therefore accept as a qualitative indication corresponding to the change caused by frictional work. We therefore ask whether there is any other phenomenon known to us which can also cause this 'happening', this rise of temperature, without friction – and we must be absolutely sure that it *is* without frictional work of any kind if we are going to identify it with ΔQ_t.

This is perhaps more difficult to do in modern times than it was a hundred years ago, because we are all so familiar with electrical appliances which we 'switch on', and observe a rise of temperature. We need some level of scientific sophistication to appreciate that this also is a case of work being done against friction. There is also the common phenomenon of combustion, which causes rise of temperature, and this has certainly been known through all man's recorded history, and it was undoubtedly this common knowledge which made it difficult for scientists in the beginnings of thermodynamics to relate temperature change clearly to energy and mechanics. However, we can dispose of electric fires and combustion quite definitely if we hold fast to the concept that what we are looking for is a *process of transmission*. In the rise of temperature of the electric appliance, and of the flame, even without any sophisticated idea of what is happening, we can certainly recognise that it is happening *internally*, and not being transmitted from somewhere else. To identify ΔQ_t we must find some phenomenon which causes rise of temperature in a reference body and which is recognisably transmission from something outside the reference body.

The second clue is that while in its very nature ΔW_f is always positive, and must always therefore give an indication in the same direction – observed to be rise of temperature – ΔQ_t is transmitted energy and hence it must include the case of transmission *from*, as well as transmission *to* the reference body. Since the result of transmission to a reference body has been identified as rise of temperature, the result of transmission from a reference body must be identifiable as fall of temperature. We must therefore also seek in our experience for conditions where a fall of temperature can be observed.

This conclusion now enables us to make an unmistakeable identification, for one of our common experiences is that when two bodies are at different temperatures θ_a and θ_b, and if, so far as we can tell, they are isolated from everything else except each other, then the lower will increase and the higher will fall. Alternatively we observe that if the higher is maintained, the lower will continue to increase until the two are equal, while if the lower is maintained, the higher will fall to equality. This common phenomenon is that which we call 'heating', and for untold centuries man has used this process to raise the temperature of his dwelling place by the effect of the higher temperature produced by combustion, and indeed has 'warmed' himself in the sunshine and 'cooled' himself in the sea.

We therefore identify the process of energy transmission, entirely distinct in character from work, which was deduced to exist, with that process which we call 'heating'. We know also that the defining characteristic of this process is that transmission can only occur when there is a *difference* of temperature. Hence this characteristic which we call temperature is involved in the matter in two quite distinct ways. The first is as a mere *indication* that the quantity $dU+p\,dV$ has changed, and the second is that a *difference* of temperature is necessary for transmission of energy by the heating process. We shall discuss temperature much more fully in later sections. Meanwhile the relative importance of these two different ways in which it arises in our study is worth comment.

In the first way in which temperature enters our discussion, i.e. as an indicator of the change $dU+p\,dV$ when work is done against friction, it is rather adventitious. The logic which led us to belief in internal energy and energy conservation would demand that in the circumstances discussed, some change must occur, even if we had no indication of it. And there are circumstances when we might not get a rise of temperature as the indication. For suppose we do the frictional work on water at atmospheric pressure and at that temperature which we arbitrarily call 100 °C. The water will boil instead of increasing in temperature. We therefore accept that evaporation, phase change, is an alternative indication of the change $dU+p\,dV$, and of course in this case the volume change from liquid to vapour is very great. But the second way in which temperature enters our discussion constitutes the vital defining characteristic of the process, i.e. that the transmission of energy by heating only occurs when a difference of temperature exists. The importance of these two aspects of temperature are discussed fully in the sections dealing with the zeroth and second laws of thermodynamics. Meantime we are now in a position to state what is called the first law, having identified ΔQ_t specifically as energy received by a body, per unit mass, via the process of heating, i.e. the process of transmission by means of a temperature difference.

Note that if ΔQ_t is negative, colloquially it would be called 'cooling', but we shall keep strictly to the single word heating as the name for the process, so that a negative ΔQ_t will be said to mean that energy is transmitted *from* the body by heating, or by the heating process.

We also recognise that by making special arrangements of particular materials we may effectively 'insulate' a reference body from all others so that ΔQ_t may be zero, in particular cases, just as it would be zero if there were no temperature difference. All processes occurring under the condition $\Delta Q_t = 0$ are classified under the special name *adiabatic*. Thus for any adiabatic process

$$dU + p\,dV = \Delta W_f,$$

and therefore

$$\Delta W_t = -p\,dV - d(\tfrac{1}{2}v^2) - d\Phi - \Delta W_f$$

$$= -d[U + \tfrac{1}{2}v^2 + \Phi] = -d\epsilon. \qquad (2.1.2)$$

This result of course could also be taken directly from equation (1.6.3), when $\Delta\Psi_t$ contains no mode other than work.

2.2 The first law of thermodynamics

It is customary to refer to *the* first law of thermodynamics as if there were only one single statement involved. In fact, however, there are several aspects of what is really meant by the first law, and this gives rise to different interpretations and some confusion.

From many points of view the basic essential 'first law' might be identified with the principle of conservation of energy. The most general statement of this principle is that summed up in equation (1.6.3) and repeated here

$$d\epsilon = \Delta\Psi_t, \qquad (2.2.1)$$

or, in words:

The change in the total energy content of a specified mass is equal to the net or total energy transmitted to it from all other masses by whatever processes of transmission may exist.

However, we may regard the first law of *thermodynamics* as giving the further information that $\Delta\Psi_t$ can occur only by *two* different processes, thermal via a temperature difference, i.e. heating, and mechanical via a force, i.e. work. This is most generally stated, from (2.2.1), as

$$d\epsilon = \Delta Q_t - \Delta W_t. \qquad (2.2.2)$$

or, in words:

The change in total energy content of a specified mass is equal to the energy transmitted to it by heating, less the net or total energy transmitted from it by work.

Since $\qquad \epsilon = U + \tfrac{1}{2}v^2 + \Phi \quad$ from (1.6.1),

and $\qquad \Delta W_t = p\,dV - d(\tfrac{1}{2}v^2) - d\Phi - \Delta W_f \quad$ from (1.4.3), and $\qquad \Delta Q_t = dU + p\,dV - \Delta W_f$ from (1.6.6),

equation (2.2.2) is satisfied, and indeed (1.6.6) was deduced from satisfying it.

When the reference mass is at rest, or is considered in relation to an environment moving uniformly with itself $\epsilon = U$, and $\Delta W_t = p\,dV - \Delta W_f$, so that equation (2.2.2) takes the alternative form

$$dU = \Delta Q_t - \Delta W_t, \tag{2.2.3}$$

or in words:

The change in internal energy content of a specified mass at rest relative to its environment is equal to the energy transmitted to it by heating, less the net or total energy transmitted from it by work.

When the reference mass moves in relation to its environment another useful equation is obtained by using the *dynamic* work ΔW_d instead of ΔW_t. We recall from (1.4.1) and (1.4.2) that

$$\Delta W_t = \Delta W_d + d(pV). \tag{2.2.4}$$

Substitution in (2.2.2) gives

$$d\epsilon = \Delta Q_t - \Delta W_d - d(pV). \tag{2.2.5}$$

$$\therefore \quad d[\epsilon + pV] = \Delta Q_t - \Delta W_d. \tag{2.2.6}$$

$$\therefore \quad d[U + pV + \tfrac{1}{2}v^2 + \Phi] = \Delta Q_t - \Delta W_d. \tag{2.2.7}$$

Now if the mass is moved, but is finally brought back to its initial velocity and potential, $d(\frac{1}{2}v^2)$ and $d\Phi$ are zero, so that for these conditions equation (2.2.7) becomes

$$d[U + pV] = \Delta Q_t - \Delta W_d. \tag{2.2.8}$$

The property sum $U + pV$ is called the *enthalpy* of the mass and is given the symbol H, so that we may express equation (2.2.8) in words as:

The change in enthalpy content of a specified mass which is moved but finally restored to its initial velocity and potential is equal to the energy transmitted to it by heating, less the energy transmitted from it by dynamic work.

By an obvious extension of definition it is usual to define the quantity $H + \frac{1}{2}v^2$ as *stagnation enthalpy*, and $H + \frac{1}{2}v^2 + \Phi$ as *stagnation and potential enthalpy* or *total enthalpy*. Equation (2.2.7) can therefore be expressed in words similar to those used in expressing equation (2.2.8).

When applied to a continuous flow of mass at a steady rate through some device, equation (2.2.7) is frequently said to be the *steady-flow energy equation.*

The important point is to note that the difference in form between the equation

$$dU = \Delta Q_t - \Delta W_t \qquad (2.2.3)$$

and

$$dH = \Delta Q_t - \Delta W_d \qquad (2.2.7) \text{ or } (2.2.8)$$

is really superficial, since in fact, by definition

$$\begin{aligned} dH &= dU + d(pV) \\ &= dU + (\Delta W_t - \Delta W_d). \end{aligned}$$

Nevertheless, for cases in which the substance is moved, the relation in terms of enthalpy and dynamic work is usually more helpful than the relation in terms of internal energy and net or total work.

It will be seen that each of equations (2.2.1), (2.2.2), (2.2.3) and (2.2.7)/(2.2.8) express the conservation of energy principle. It may be asked whether it is justifiable to take any one of them as more worthy than the others of being chosen as *the* first law of thermodynamics. On the whole this is doubtful.

However, if we concentrate on the process of transmission by heating, we have the advantage that the relation for it is known to be true whether the reference body moves or is always at rest in relation to its environment, i.e. for the heating process

$$\Delta Q_t = dU + p\,dV - \Delta W_f \qquad (2.2.9)$$

is unique and free of the complications of different forms ΔW_d, ΔW_{tp}, ΔW_t which occur in the work process. Hence there is some good reason to regard (2.2.9) as expressing the most general aspect. We shall adopt this and state the first law of thermodynamics in the following words:

In the heating mode of transmission of energy, the amount of energy transmitted to a specified mass is given by the increase of its internal energy and the work done by it in volume change against the ambient pressure, less the amount of frictional work done by it.

Thus we regard equation (2.2.9) as the mathematical expression of the first law. Many texts give (2.2.3) as the first law, but it should be noted that this is a less general statement than (2.2.9) although it coincides with (2.2.9), since for the case of no relative motion to which it refers, $\Delta W_t = p\,dV - \Delta W_f$.

2.3 Temperature. The zeroth law of thermodynamics

We have mentioned temperature as a phenomenon apparent to our senses. The position of our argument is that change of temperature shares with change of phase, the status of being adventitious evidence for the occurrence of internal energy change. The logic of our background belief in the conservation of energy would compel us to postulate the existence of internal energy as a variable quantity even if we had no sensible evidence of its variation. We have also established that what we call the heating process transfers energy to a portion of recipient substance. Now in the heating process as we experience it, the changes of temperature and/or phase which we identify with energy content change can be made to occur only in one governing circumstance. We find by careful observation that this circumstance is a difference of the kind which we call hotter/colder, i.e. a difference of temperature itself. Thus we realise that this property 'temperature' quite apart from its adventitious characteristic of reflecting *changes* of internal energy (provided only a single phase is present) is somehow inherently related to the process of energy *ex*change by heating.

One of the basic semantic/logical problems in thermodynamics is how we make the next step from there. Qualitatively we have a sense of a warmer/colder relationship. But in the well-established interpretative observation that energy exchange by heating only occurs when there is a warmer/colder difference, we have an element of the 'chicken and egg' sequence problem. True it is that difference is always observed when energy exchange by heating occurs, but it is equally true that energy exchange by heating will also occur whenever there is warmer/colder difference. (Insulation merely slows down the rate of exchange.) Given that energy exchange by heating is occurring, is it a mere tautology to add the statement that the warmer/colder difference is also present, or is it a statement which has a significant meaning distinct from the statement about energy exchange? The *zeroth law* of thermodynamics is essentially the postulate that the warmer/colder difference statement has a distinct and significant meaning despite the fact that it inevitably occurs in association with energy exchange by heating. We contemplate a possible limit situation when no energy exchange by heating would be found to occur, and then assert that in such a situation no part could be said to be warmer nor colder than any other, i.e. it could then be said that a single *value* of some measure of the 'warmth' exists. In brief the zeroth law is the postulate that it is meaningful to talk about the temperature of a piece of matter, defined as being the value of such measure.

We do not need at this stage in our logic to be specific about how a numerical value is to be assigned to the temperature. All we are concerned with is to establish that it is a meaningful concept, so that later, when other aspects have become known, we may then discuss how a numerical value may be given to it. The zeroth law – so named because the names *first* and *second* laws had already become accepted for other basic laws of the subject before it was thought necessary to regard the very existence of the property temperature as an essential prior postulate – refers to conditions where energy exchange by heating is not occurring. We now pass to the second law which deals with the occurrence of such exchange.

2.4 Temperature difference. The second law of thermodynamics

This law is the most simple of the three to state – at least this is so after the discussion of the zeroth law. For having decided that it is not mere tautology to say that energy transfer by heating only occurs when a difference of temperature is present, we recognise that the temperature difference and the energy transfer are distinct variables, with a functional relation between them. All that is now necessary is to examine the characteristics of the functional relationship as revealed in our experience and state the most general truth we can observe about it. We find that all our experience can be covered within the following statements.

(*a*) The amounts of energy transferred depend upon the kinds of material, their relative motion, the time duration, and on the temperature difference.

(*b*) The dependence of the amount transferred per unit time on very many physical factors as well as temperature difference requires a whole detailed study in itself, and there appears no hope of a simple *quantitative* generalisation.

(*c*) Nevertheless one simple qualitative generalisation is possible. This can be stated by reference to two masses between which transfer of energy by heating takes place. It is found that the recipient substance, defined as that for which $m_a[dU_a + p_a\, dV_a - \Delta W_{fa}]$ is positive, *always* has the lower of the two temperatures. The donor substance for which $m_b[dU_b + p_b\, dV_b - \Delta W_{jb}]$ has the same arithmetic value but is negative, is always at the higher temperature. We know of no instance in our experience of the process we call heating which does not fall within this generalisation.

Thus we formulate the second law of thermodynamics as follows.

'*In the process of energy transmission by heating, the energy so transferred goes from substance at higher temperature to substance at lower temperature and never from the lower to the higher.*'

Various formulations of the second law exist, all of which are equivalent to the above. One point of comment may be helpful, however. It is well-known that by using a machine (a refrigerator) energy may be taken from a substance at a lower temperature and ultimately transferred to one at a high temperature. Thus the formulation given above might have been expected to include the caveat 'without the assistance of a work machine' or 'of itself', or some other such qualification. Many formulations of the second law do include this kind of clause to cope with the fact of the possibility of refrigeration. But we believe this to be unsound. In any refrigerator, the energy taken from the reference substance at a lower temperature θ_s for delivery to the other reference substance at a higher temperature θ_r must *always* primarily be transferred by heating from θ_s to an intermediate substance at a temperature θ_{ss} lower than θ_s, and the intermediate substance has to be raised to a temperature θ_{rr} *higher* than θ_r by the work process, before energy can be transferred by heating to θ_r.

Hence the formulation given is to be regarded as general, and complete, and we believe it to be universally true.

2.5 The laws for real processes

We now examine in retrospect the three laws of thermodynamics, zeroth, first, and second, and consider if, and how, they may be regarded as applicable to actual processes occurring in the real world.

The discussion given in Section 2.3 shows that the phenomena of energy exchange by heating and of a warmer/colder relationship always occur together. This co-occurrence constitutes both preamble to the zeroth law and the basis of the second law. The latter adds merely a statement of *direction* of transmission. This is indeed an important addition for its practical consequences, but in regard to the logical position the description 'merely' is appropriate. No difficulty or query arises in the addition of an observation of direction. But the zeroth law is strictly established by assuming that a condition of vanishing energy exchange by heating will also lead to a vanishing of any warmer/colder relationship, and hence the concept of temperature is established. It follows that when any real finite portion of substance is undergoing a real process in which finite energy transmission by heating occurs to it in a finite time then it cannot strictly be said to have *a* temperature. Conventionally we overcome this difficulty by adopting the notions of the differential calculus, and *assume* that the concept of temperature is nevertheless valid with reference to an infinitesimal portion of substance, even if finite energy exchange to and from the infinitesimal portion is occurring. On this assumption the concept of temperature *gradient* is also justified, and we visualise energy transmission by heating to take place *down* any temperature gradient, in accordance with the second law.

Clearly these assumptions run into logical difficulties when the particulate nature of matter is recalled. If our 'infinitesimal portion' of a gas contains only one molecule, what can we mean by 'temperature', or 'energy exchange by heating'? This difficulty is not of course unique to 'temperature', because for a single molecule we have equal difficulty in conceiving what could be meant by its 'pressure', or 'volume' or 'energy exchange by work'. Such difficulties can only be dealt with properly by recourse to statistical thermodynamics. For the present purposes of continuum thermodynamics we ignore the difficulties, by postulating that the infinitesimal element is nevertheless large enough to contain many molecules. Since a cubic millimetre of a gas at normal temperature and pressure contains about 10^{16} molecules, the condition is reasonably satisfied under most practical conditions.

On this basis we also assume that in any finite portion of substance undergoing a real process it is also legitimate to accept the concept of local values of pressure, density, internal energy, velocity, etc., together with the associated idea of corresponding gradients. Such characteristics will only be uniform, with all gradients zero, throughout a finite mass, if no process is happening, and it is then said to be in equilibrium.

With this conceptual pattern we see that the second law of thermodynamics, unchanged from the form enunciated in Section 2.4, is still applicable throughout any finite size of portion of matter, while the notion of temperature used in that enunciation retains validity from the zeroth law as applied to any small infinitesimal portion. The qualitative aspects of the first law, i.e. the statements of existence of energy transmission by heating and by work, also retain validity within that framework, but what happens to the quantitative expressions?

As a preliminary to answering this question we require to give some further consideration to the equilibrium condition of uniform pressure and density throughout a finite mass.

2.6 Properties and equation of state

Once the process by which ΔQ_t occurs has been identified, it is possible to design experiments in which energy transmission by this process will be negligible. This can be done by making all temperatures of interacting elements in the experiment nearly equal, and by isolating them from any objects at greatly different temperatures. When we are sure that ΔQ_t is zero we then have

$$\Delta W_f = dU + p\,dV, \tag{2.6.1}$$

and at constant pressure therefore

$$W_f - p(V_2 - V_1) = U_2 - U_1. \tag{2.6.2}$$

In principle it is easy to make mechanical measurements of the terms on the L.H.S. of equation (2.6.2) and hence we can determine $U_2 - U_1$. This is what the famous Joule experiments did. A known increase in internal energy was given to a known mass of a known substance, and related to the observed rise of temperature.

An alternative approach is to try to make ΔW_f negligible as well as ΔQ_t. If both are zero we have, from (2.6.1)

$$-\int_{V_1}^{V_2} p\,dV = U_2 - U_1. \tag{2.6.3}$$

Again in principle the left-hand side may be mechanically determined, and so a known change in internal energy can be related to an observed change in temperature. This approach was also attempted by early experimenters, notably Mayer.

More sophisticated techniques of greater precision are now available, but the essential principle remains the same. As a result of such work the internal energy change of standard substances has been precisely related to standard – though arbitrary – temperature intervals and phase changes. We can therefore in principle measure also, by such calibrations, the amount of ΔQ_t which occurs in cases where energy transfer by heating is permitted. Our immediate concern here is not with the precision of such measurements but with the fact that they are possible in principle, and with the general facts which they reveal. And one aspect arises which requires special comment – this concerns why it was necessary to refer earlier in this paragraph to 'standard substances'.

When we are considering the Newtonian laws of mechanics for a particle, we have no concern whatever about what 'kind' of substance is in the particle. The question of 'kind' just does not arise. But as soon as we go, as we did in Section 1.2 of Chapter 1 and have continued onwards, to a finite piece of matter, however small, the *kind* of substance is implicit. We have referred so far to our basic experiences of mechanical phenomena and thermal phenomena. But there is a third experience of which we are all equally aware from our earliest days. Not only do things move and give us the sensations of force and of warmth, but they also appear different in kind. Water, iron, wood, stone, clothing, and all the things we see, exhibit immense variety of kind, i.e. in scientific terms we are aware of three classes of phenomena; mechanical, thermal, and chemical. Now in the equations which we developed in Chapter 1, thermal and chemical characteristics were actually implied, without comment, since the equations include the specific volume V of the substance, and we know from observation that the density of a substance depends not only on the mechanical condition summarised in the stress p, but also on the thermal characteristic we call temperature and on the chemical kind of material.

It is therefore not surprising that, when experimental investigations are undertaken, as exemplified at the beginning of this section, to determine the changes of temperature and phase associated with change of internal energy, it is found that the amount of energy supplied to cause a definite temperature change or phase change to occur in a unit of mass depends on the kind of substance, the chemical nature of the unit of mass, and on the initial temperature. Thus for example it takes more energy to raise the temperature of a kilogram of water by one of our arbitrary temperature units than it does to cause the same rise in a kilogram of iron, and in each substance it takes differing amounts of energy to cause the same rise starting from differing initial temperatures. Similarly the energies required to cause evaporation or freezing of the substances are different.

Thus in all our further discussions the presence of the internal energy U will reinforce the specific volume V as a reminder that some particular chemical substance is also implied in what we are doing. This is a very important aspect of thermodynamics and we shall return to it in much greater detail in a later chapter. Meantime we proceed on the assumption that we are discussing something which we may call an 'inactive' substance, i.e. which can be regarded as consisting of something in which no chemical change occurs, as for example, only one single chemical species. We accept this as a qualitative idea meantime, and obtain later a more precise idea of what it means.

Thus it is a fact of experience that the volume occupied by a given mass depends on *three* other items which have to be known, viz.

(*a*) the chemical nature of the substance;
(*b*) its state of stress;
(*c*) its temperature and/or phase (solid, liquid, gas).

We recall here that at present 'temperature' and 'change of phase' have the same status as adventitious, but useful, indications of change of internal energy. Thus (*c*) really means that specific volume depends on internal energy as well as on (*a*) and (*b*).

For our present purposes it is enough to concentrate on (*b*) and (*c*) only, i.e. on the implications which arise from the dependence of specific volume of a given substance on its state of stress and on its internal energy. With solids the discussion is complicated by the existence of unequal stress in different directions. We limit our discussion to fluids for which the pressure, uniform in all directions, is the only relevant stress.

This general relation can be expressed in mathematical form by saying that a single fluid substance has an equation of state of the form

$$p = p(U, V),$$

or

$$U = U(p, V),$$

or

$$V = V(p, U). \tag{2.6.4}$$

Each of the variables p, U and V is expressible in the fundamental mechanical terms, based ultimately on mass, length, and time, which are consistent with the logic of our approach.

Anticipating further development, however, let us assume temporarily that we have reached *some* method of *measuring*, and thus giving quantitative values to, the phenomenon of temperature. (There is no logical need for anything other than arbitrary units – since units of mass, length and time are themselves *arbitrary* – but a principle of measurement of values has to be assumed.) Let the value of temperature then be denoted by θ. Thus, provided only a single phase is present, the equation of state (2.6.4) can be transformed into a relation between p, V and θ, i.e.

$$\left.\begin{aligned} p &= p(\theta, V), \\ \theta &= \theta(p, V), \\ V &= V(p, \theta). \end{aligned}\right\} \tag{2.6.5}$$

Equations (2.6.4) will apply always for a single substance. In the case of a single phase we may use either the form of (2.6.4) or of (2.6.5) knowing that if we use (2.6.4), temperature is also determined, and using (2.6.5), internal energy is also determined. But (2.6.4) is logically prior.

These properties p, V, U and θ which are related by the equation of state are called *thermodynamic* properties of the substance. Other thermodynamic properties exist, as for example the flow function pV and the enthalpy $H = U+pV$ which we have already noted. Some others will be found in later chapters. But all are implicitly determined by the equation of state.

Throughout this section we have been considering a finite mass of substance undergoing no process and therefore with uniform properties throughout every infinitesimal element of mass, i.e. a finite mass in equilibrium. It is to this that the equation of state strictly applies. Having established its existence for the equilibrium condition, we can now return to the point left unresolved at the end of Section 2.5, i.e. how to express the quantitative significance of the first law for a finite mass undergoing a real process and hence inherently not in strict equilibrium.

2.7 The first law for finite processes. Quasi-equilibrium condition and properties

In a real fluid process there must be at any stage some non-uniformity of properties. Local values of pressure, velocity, temperature, density, etc., will fluctuate and there will also in general be a spacial variation of the mean local values. For the present discussion we shall allow that all such non-uniformities may be present. Let us consider, at any stage in a process – as, for example, at a normal cross section of flow through a device – a representative sample of mass M of the material.

We may say that the sample mass M consists instantaneously of masses $M_1, M_2, \ldots, M_i$, etc., having properties $p_1, \ldots, p_i$, $V_1, \ldots, V_i$, $U_1, \ldots, U_i$, etc., and velocities $v_1, \ldots, v_i$. The only requirement is that each mass M_i is sufficiently small to permit the statement that it has p_i, V_i, v_i, etc. In the limit the individual masses become infinitesimal, but since we have no spacial functional relations to consider, the summation formulation may be retained and is more convenient. The fact that the masses may split up and interpenetrate subsequently need not concern us, since we can conceptually retain their identity.

Let us denote the component velocities of a mass M_i by v_{ix}, v_{iy}, v_{iz}, and then define velocities $\bar{v}_x$, $\bar{v}_y$, $\bar{v}_z$ by the following equations:

$$\left.\begin{aligned} \sum_i M_i v_{ix} &= M\bar{v}_x, \\ \sum_i M_i v_{iy} &= M\bar{v}_y, \\ \sum_i M_i v_{iz} &= M\bar{v}_z, \end{aligned}\right\} \qquad (2.7.1)$$

i.e. $\bar{v}_x$, $\bar{v}_y$, $\bar{v}_z$ are the components for which the total mass would have the same total momentum as the sum of momenta of all the individual masses in the same direction, i.e. $\bar{v}_x$, $\bar{v}_y$, $\bar{v}_z$ constitute the component velocities of the centre of mass of M.

Now define
$$\left.\begin{aligned} v_{rix} &= v_{ix}-\bar{v}_x, \\ v_{riy} &= v_{iy}-\bar{v}_y, \\ v_{riz} &= v_{iz}-\bar{v}_z, \end{aligned}\right\} \qquad (2.7.2)$$
and hence we find
$$\sum_i M_i \frac{v_{ix}^2}{2} = M\frac{\bar{v}_x^2}{2}+\sum_i \frac{v_{rix}^2}{2}+\bar{v}_x \sum_i M_i v_{rix}.$$
But
$$\sum_i M_i v_{rix} = 0,$$
$$\therefore \quad \sum_i M_i \frac{v_{ix}^2}{2} = M\frac{\bar{v}_x^2}{2}+\sum_i M_i \frac{v_{rix}^2}{2}. \qquad (2.7.3)$$

Using the corresponding result for the other two components, and summing all three, gives finally
$$\sum_i M_i \frac{v_i^2}{2} = M\frac{\bar{v}^2}{2}+\sum_i M_i \frac{v_{ri}^2}{2}, \qquad (2.7.4)$$
i.e. the total kinetic energy is the sum of the kinetic energy of the total mass in respect of the motion of its centre of mass, plus the sum of kinetic energies due to the motion of the individual masses relative to the centre of mass of the whole sample.

The total energy content of the mass M is therefore
$$\begin{aligned} \epsilon' &= \sum_i M_i \left(U_i+\frac{v_i^2}{2}+\Phi_i\right) \\ &= \sum_i M_i U_i+\sum_i M_i \frac{v_{ri}^2}{2}+M\frac{\bar{v}^2}{2}+M\bar{\Phi}. \end{aligned} \qquad (2.7.5)$$

Here we have written $\quad \sum_i M_i \Phi_i = M\bar{\Phi}, \qquad (2.7.6)$

which effectively defines $\bar{\Phi}$ as the potential per unit mass of M considered at the position of its centre of mass.

Hence if the sample mass M is large enough to be representative of the condition of the substance at one particular stage of the process, we can say that unit mass of the substance at that stage has an energy content
$$\epsilon = \sum_i \frac{M_i U_i}{M}+\sum_i \frac{M_i \frac{v_{ri}^2}{2}}{M}+\frac{\bar{v}^2}{2}+\bar{\Phi}. \qquad (2.7.7)$$

The quantity $\sum_i (M_i U_i/M)$ is evidently the mean internal energy of the substance per unit mass, and may be denoted by $\overline{U}$.

Similarly we can say that the mean specific volume of the substance is

$$\overline{V} = \sum_i \frac{M_i V_i}{M}. \tag{2.7.8}$$

Now there will exist for the substance a uniform equilibrium state having internal energy U^* and specific volume V^* such that $U^* = \overline{U}$ and $V^* = \overline{V}$, the asterisk denoting in each case the properties of a true uniform equilibrium state. The specifications of U^* and V^* are sufficient to determine the pressure p^* and all other properties from the equation of state. The same equation of state necessarily relates p_i, V_i, and U_i in each subordinate mass in the non-uniform case. Because of this it can be shown that in general for real substances the summation $\sum_i M_i p_i V_i/M$ cannot be equal to p^*V^* when $U^* = \overline{U}$ and $V^* = \overline{V}$. Hence although we can postulate an equilibrium state matching the actual non-uniform condition in internal energy and specific volume, there cannot be a match in enthalpy.

Thus when a substance is undergoing a real process, while we may consider it kinematically as moving with a mean velocity $\bar{v}$ and having a mean potential $\overline{\Phi}$, and thermodynamically as having mean specific internal energy $\overline{U}$ and specific volume $\overline{V}$, the condition does not correspond exactly to an ideal uniform stream. The kinematic fluctuations give rise to the additional kinetic energy term

$$\sum_i M_i \left(\frac{v_{ri}^2}{2}\right) \Big/ M$$

and the thermodynamic fluctuations give rise to the difference between

$$\sum_i M_i p_i V_i \quad \text{and} \quad M p^* V^*.$$

Nevertheless, the concept of an equivalent uniform condition is exceedingly useful and important.

The crucial matter is the size and scale of the non-uniformities in the actual condition. Just because there is non-uniformity, we could in principle get work output if we use a device which is small enough to get 'in between' the non-uniformities. But if the velocity differences, and the relative motions and the pressure and density variations, are so closely spaced that any device we can make cannot be sufficiently small to get in between the non-uniformities, we shall not in fact be able to make any use of them. Clearly such extraction from non-uniformities can be done when the size range is sufficiently large. We have mechanisms – windmills – which obtain work from the variation in air pressure and velocity in the atmosphere. A glider can get lift from convection currents. We make use of systematic non-uniformities as, for example, such as occur with circulation in hot water systems and boilers. In all such cases the scale of the non-uniformity is at least of the same order as the relevant equipment.

But in thermodynamics we are concerned with the processes of energy interaction as a whole, at least so far as the general laws are concerned. The question of relative size is fundamental. As Clerk Maxwell pointed out long ago, if a small enough and fast enough device could be operated to separate out the fast moving and the slow moving molecules in a gas we should be able to get work from that. The same is true of the random energy in turbulent flow and of all other fluctuating flow behaviour. But for general thermodynamic purposes we exclude the possibility of getting energy transmission by work from the non-uniformities and fluctuations themselves. In the flow of fluids in engines, compressors, pumps, turbines, fans, etc., the scale of turbulence and fluctuations is very small compared with the size of the operating parts of the machines.

Subject to this restriction – i.e. to the exclusion of the possibility of obtaining work transmission from the non-uniformities – the representation of the actual non-uniform condition of a fluid in a real process as equivalent to the same substance in an equilibrium state $U^* = \overline{U}$, $V^* = \overline{V}$ flowing with uniform velocity $v^* = \overline{v}$ is justified.† Hence we can apply the first law using the equations of Section 2.2 of this chapter in the corresponding forms:

$$dU^* = \Delta\overline{Q}_t - \Delta\overline{W}_t, \qquad (2.7.9)$$

$$dH^* = \Delta\overline{Q}_t - \Delta\overline{W}_d, \qquad (2.7.10)$$

$$\Delta\overline{Q}_t = dU^* + p^*\,dV^* - \Delta\overline{W}_f. \qquad (2.7.11)$$

Similarly the relations from Chapter 1 become:

$$\Delta\overline{W}_t = p^*\,dV^* - d(\tfrac{1}{2}\overline{v}^2) - d\overline{\Phi} - \Delta\overline{W}_f, \qquad (2.7.12)$$

$$\Delta\overline{W}_d = -V^*\,dp^* - d(\tfrac{1}{2}\overline{v}^2) - d\overline{\Phi} - \Delta\overline{W}_f. \qquad (2.7.13)$$

† A detailed analytic justification follows in Section 2.9 of this chapter. This may be omitted by the beginner.

The properties U^*, V^* and p^*, H^*, θ^* which then follow from the equation of state, for the equilibrium state determined by $U^* = \bar{U}$ and $V^* = \bar{V}$, are commonly called quasi-equilibrium properties, and the state itself is referred to frequently as the quasi-equilibrium state. The bar sign on the energy transmissions by heating and by work, and on the velocity and potential, denote averages for a finite mass undergoing a real process.

With this formalism we can now accept the three laws of thermodynamics, zeroth, first, and second, as fully applicable to real processes, and proceed to discuss actual processes in these terms. For simplicity in nomenclature we shall henceforth omit the asterisk sign on the quasi-equilibrium properties and the bar sign on the average velocity, potential and transmissions, and use simply U, V, p, θ, $\frac{1}{2}v^2$, Φ, ΔW_t, ΔW_d, ΔQ_t, etc. It will be understood however that – except under very special conditions where uniformity might occur, and which will be appropriately mentioned in the context – we shall always be referring to real processes and quasi-equilibrium.

2.8 Process and thermodynamic path. A major consequence of the first law

Let us consider for a particular substance a graph wherein the abscissa is V and the ordinate is p, as shown in Figure 2.8.1. Then from the discussion in Section 2.6 any point in the graph represents a single state of the substance, i.e. each point is a definite unique state whereat U, as well as p and V, is defined. (For a real process it is of course quasi-equilibrium which is implied.)

A curve such as *abc* on the graph represents a sequence of such unique states, and the transition of the substance from state (a) to state (c) by the curve *abc* represents a *particular* process. Going by the curve *ab′c* represents quite a different process, but it begins and finishes with the substance at the same state, i.e. the property values p_c, V_c, U_c are the same whether we arrive at c via b or via b'. Now $\int_a^c dU$ is simply $U_c - U_a$ and again is independent of whether the process reaches c via b or b'. But

$$\int_a^c p\,dV = \text{Area } abcdea \quad \text{if we proceed via } b, \tag{2.8.1}$$

and

$$\int_a^c p\,dV = \text{Area } ab'cdea \quad \text{if we proceed via } b'. \tag{2.8.2}$$

Hence $Q_{t,abc}$ is different from $Q_{t,ab'c}$, since in general

$$Q_t = \int dU + \int p\,dV - \int \Delta W_f. \tag{2.8.3}$$

Similarly $W_{t,abc}$ is different from $W_{t,ab'c}$ and $W_{d,abc}$ differs from $W_{d,ab'c}$ because of the presence of the integrals $p\,dV$ and $-\int V\,dp$ which depend upon which path is followed.

This now reveals much more clearly what is meant by saying that neither the energy transmitted by heating nor that transmitted by working are properties. They occur in processes, by which the values of

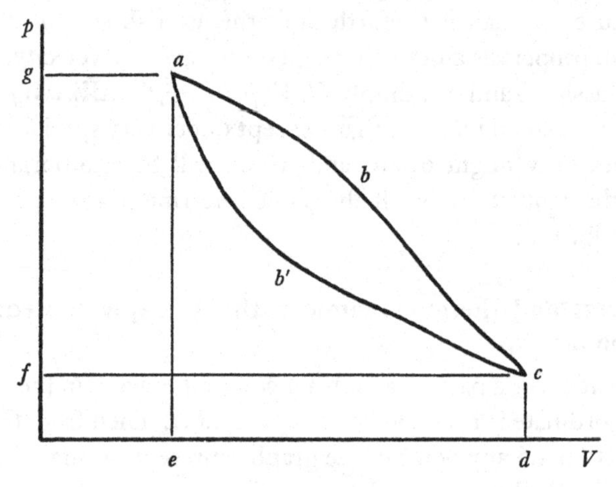

Fig. 2.8.1

properties may be changed. Thus we cannot specify the amount of energy transferred by either heating or working without specifying the process path, and the amount of frictional resistance.

We shall adopt the notation $[b\int_a^c$ to indicate an integral taken along a particular path b. From equation (2.8.3) we have therefore

$$Q_{t,b} = [b\int_a^c dU + [b\int_a^c p\,dV - [b\int_a^c \Delta W_f, \tag{2.8.4}$$

or

$$Q_{t,b} + [b\int_a^c \Delta W_f = [b\int_a^c dU + [b\int_a^c p\,dV. \tag{2.8.5}$$

Now as we have seen, once a particular path and its end points are stated, *both* $\int dU$ and $\int p\,dV$ are completely determined, although the former only needs the end points for its determination.

Thus equation (2.8.5) tells us that for a particular thermodynamic path between two given end points, the *sum* of the energy transmitted to the substance by heating *and* of that dissipated against friction is constant. Thus we can in principle proceed along the same path either with a large intake of energy by heating and a small dissipation against friction or by smaller absorption of energy by heating and larger frictional dissipation.

This is a major consequence of the first law for real processes. Since ΔW_f must be positive, the energy intake by heating to follow any particular thermodynamic path is largest when we have the smallest frictional resistance. The maximum energy intake by heating for a particular path is therefore that which would be obtained in an ideal frictionless mechanically reversible process. If the path is one in which energy is rejected by heating instead of being absorbed, i.e. if $Q_{t,b}$ is negative, then the same theorem implies that the *minimum* energy rejection by heating for a particular path is that which would be obtained in an ideal frictionless process.

The important point from the above discussion is that we can in principle move along the *same* path, with the same $\int(dU+p\,dV)$ by an infinite range of processes so long as the sum $Q_{t,b}+[b\int_a^c \Delta W_f$ (and of course each element sum $\Delta Q_t+\Delta W_f$) is constant. If we have high friction we cannot put in so much energy by heating, or we have to reject more energy by heating, to follow a specified path. This is the basic importance of the first law statement,

$$\Delta Q_t = dU+p\,dV-\Delta W_f.$$

This importance may be brought to our attention more forcibly by rewriting it in the form

$$\Delta Q_t+\Delta W_f = dU+p\,dV. \qquad (2.8.6)$$

We may note that while we have discussed the behaviour of ΔQ_t in processes explicitly in terms of the first law only, the zeroth and second laws are also implied. Unless ΔQ_t is zero, there must be available some other body, which must always be at a higher temperature than that of the reference substance if ΔQ_t is positive, and lower if ΔQ_t is negative. If such temperature difference were vanishingly small so that ΔQ_t would be transmitted only very slowly, the process would be in principle *thermally* reversible. This concept of thermal reversibility is quite distinct from that of mechanical reversibility in the ideal absence of friction. Both thermal reversibility and mechanical reversibility are unattainable ideals. In principle we can imagine either ideal to apply separately or that both apply together as an ultimate ideal. This latter is called complete thermodynamic reversibility. Under these ultimate ideal conditions we should have true equilibrium properties at every point in the path. Under either mechanically reversible or thermally reversible conditions separately or in the real process which is both mechanically and thermally irreversible, true equilibrium does not exist throughout a finite mass and quasi-equilibrium properties are implied. Nevertheless the first law has the definite quantitative consequences discussed above as a result of mechanical irreversibility. We have to reach a later stage before we can find a *quantitative* consequence of the second law as a result of thermal irreversibility.

Meanwhile it should be noted that although some thermodynamics texts will be found to refer to reversible and irreversible *paths*, it is not really the *path* which is reversible or irreversible – except for the presence of quasi-equilibrium instead of true equilibrium. It is the *process* which is reversible or irreversible, in practice always irreversible. As shown above, we can in principle follow the same path by an infinite range of processes and in particular can make a real process follow a path identical to that followed by a reversible process provided we transmit to the reference substance less energy by heating. For a reversible process, the energy transmitted by heating along a particular path b will be

$$[b\int_a^c (dU + p\,dV).$$

For a real process it will be

$$[b\int_a^c (dU + p\,dV) - [b\int_a^c \Delta W_f,$$

i.e. the integrals of dU and $p\,dV$ evaluated *along the same path*, less the frictional dissipation. These last three sentences are mere repetition of what was said a few paragraphs back, but the point that irreversibility is a characteristic of a process and *not* of a thermodynamic path is of sufficient importance to deserve repetitive emphasis.

This concludes the main line of discussion of this chapter on the laws of thermodynamics and their most immediate logical consequences. We proceed in Chapter 3 to examine their implications on the utility of energy sources – the production of power.

2.9 Appendix

In this section we take the basic description of the real non-uniform process as set out in Section 2.7, but develop the analysis to prove the validity of using the quasi-equilibrium properties.

We may apply the dynamic work equation (1.3.16) to the motion of any element mass M_i and obtain

$$\Delta W'_{d,i} = -M_i[d(\tfrac{1}{2}v_i^2) + d\Phi_i + V_i\,dp_i + \Delta W_{f,i}]. \qquad (2.9.1)$$

This accounts for dynamic work energy transmission from M_i to any moving object in contact with it, including other elements M_f, M_k, etc. When summed over all M_i the mutual interactions must cancel and the result must give the energy transmission from M as a whole by dynamic work to any moving device through which the substance is moving. Thus we have, for the mean dynamic work per unit mass

$$\Delta \overline{W}_d = \frac{\sum_i \Delta W'_{d,i}}{M} = -\sum_i \frac{M_i[d(\tfrac{1}{2}v_i^2) + d\Phi_i + V_i\,dp_i + \Delta W_{f,i}]}{M}. \qquad (2.9.2)$$

Using equations (2.7.4) and (2.7.6) gives

$$\Delta \overline{W}_d = -\sum_i \frac{M_i[V_i\,dp_i + d(\tfrac{1}{2}v_{ri}^2) + \Delta W_{f,i}]}{M} - d(\tfrac{1}{2}\overline{v}^2) - d\overline{\Phi}. \qquad (2.9.3)$$

Now we may write

$$V_i\,dp_i = V_i\,dp^* - V_i\,d(p^* - p_i), \qquad (2.9.4)$$

$$\therefore \quad \sum_i \frac{M_i V_i dp_i}{M} = V^*\,dp^* + \sum_i \frac{M_i V_i d(p_i - p^*)}{M}. \qquad (2.9.5)$$

Substitution in (2.9.3) gives

$$\Delta \overline{W}_d = -V^*\,dp^* - \sum_i M_i \frac{[V_i d(p_i - p^*) + d(\tfrac{1}{2}v_{ri}^2) + \Delta W_{f,i}]}{M} - d(\tfrac{1}{2}\overline{v}^2) - d\overline{\Phi}. \qquad (2.9.6)$$

Comparing equation (2.9.6) with equation (1.4.4), we observe that they would be identical in form if we defined a mean frictional work $\Delta\overline{W}_f$ per unit mass by the relation

$$M\Delta\overline{W}_f = \sum_i M_i[V_i\,d(p_i - p^*) + d(\tfrac{1}{2}v_{ri}^2) + \Delta W_{f,i}]. \qquad (2.9.7)$$

Now from its nature $\sum_i M_i \Delta W_{f,i}$ must cover all mutual frictional work action between the various subordinate masses, and also the net frictional work action between all of them and any device through which the substance is moving. But the net frictional work action between the whole mass M and the device must correspond to what we mean by $M\Delta\overline{W}_f$. Hence we may write

$$\sum_i M_i \Delta W_{f,i} = M\Delta\overline{W}_f + \sum_i M_i \Delta W_{f,r,i}. \tag{2.9.8}$$

From (2.9.7) and (2.9.8) we obtain

$$\sum_i M_i [V_i d(p_i - p^*) + d(\tfrac{1}{2}v_{ri}^2) + \Delta W_{f,r,i}] = 0, \tag{2.9.9}$$

where $$\sum_i M_i \Delta W_{f,r,i} = \text{sum of } \textit{mutual} \text{ frictional work.} \tag{2.9.10}$$

Equation (2.9.9) shows that over all the subordinate masses the pressure difference $p_i - p^*$ *relative* to the mean behaves in the ordinary Bernoulli equation form to give the relative kinetic energy and overcome mutual friction. Then equation (2.9.9) with (2.9.8) allows us to write (2.9.6) precisely in the same form as (1.4.4), viz.

$$\Delta\overline{W}_d = -V^* dp^* - d(\tfrac{1}{2}\bar{v}^2) - d\overline{\Phi} - \Delta\overline{W}_f. \tag{2.9.11}$$

Now just as in Chapter 1 we considered the transport work in the infinitesimal flow stream, we proceed to consider the transport work in the real, finite, non-uniform case. The transport work done by the elemental mass M_i is $M_i d(p_i V_i)$ and the total transport work for the finite mass M is formally $\sum_i M_i d(p_i V_i)$. However, on the assumption, already discussed, that the device for work transmission is large compared with the scale of non-uniformities these individual contributions cannot wholly be realised as *transmitted* work. To some extent they must be effective only in transport of elements of mass merely relative to each other, and such relative transport will contribute nothing to transmission via a device which is too large to get 'in between' the elements. Now we shall make the reasonable assumption that the amount of relative transport work can be assessed by considering each element mass M_i with a flow function $p_i V_i$ relative to the mean equilibrium flow function $p^* V^*$ and hence take the total relative transport work to be $\sum_i M_i d(p_i V_i - p^* V^*)$.

On this assumption the effective transmitted transport work from the mass M is therefore simply $\sum_i M_i d(p^* V^*)$ which is of course $M\, d(p^* V^*)$. Hence we obtain

$$\Delta\overline{W}_{tp} = d(p^* V^*). \tag{2.9.12}$$

Now as in Section 1.4 we add (2.9.12) to (2.9.11) to obtain the net or total work per unit mass in the finite non-uniform case as

$$\Delta\overline{W}_t = p^* dV^* - d(\tfrac{1}{2}\bar{v}^2) - d\bar{\Phi} - \Delta\overline{W}_f. \qquad (2.9.13)$$

It is seen that (2.9.13) and (2.9.11) now correspond exactly to (1.4.3) and (1.4.4), i.e. both $\Delta\overline{W}_t$ and $\Delta\overline{W}_d$ in the finite non-uniform case may be expressed exactly as for the infinitesimal uniform case, provided we use the mean flow velocity $\bar{v}$ and the mean potential $\bar{\Phi}$ in place of v and Φ for the infinitesimal case, and where p^*, V^* are properties of an equilibrium thermodynamic state of the substance defined by the conditions

$$U^* = \overline{U},$$

and

$$V^* = \overline{V},$$

the remaining properties, including p^*, being determined from the equation of state.

The substance is not of course actually 'in' this equilibrium thermodynamic state. It is in a non-uniform condition characterised by having average specific internal energy and average specific volume equal to the corresponding properties in the equilibrium state.

Finally, we proceed as in Chapter 1, Section 1.6, to apply the conservation of energy principle to the energy content equation (2.7.7). We have,

$$d\epsilon = dU^* + d(\tfrac{1}{2}\bar{v}^2) + d\bar{\Phi} + \sum_i M_i d(\tfrac{1}{2}v_{ri}^2)/M. \qquad (2.9.14)$$

Assuming now that, as established in Chapter 1, we may take energy transmission by heating ΔQ_t to be the only process of energy transmission other than the net or total work ΔW_t, we shall have the conservation equation as

$$\Delta\overline{Q}_t - \Delta\overline{W}_t = d\epsilon. \qquad (2.9.15)$$

Substitution of (2.9.14) and (2.9.13) in (2.9.19) gives

$$\Delta\overline{Q}_t = dU^* + p^* dV^* - \Delta\overline{W}_f + \sum_i M_i d(\tfrac{1}{2}v_{ri}^2)/M. \qquad (2.9.16)$$

In the trio of equations (2.9.13), (2.9.11) and (2.9.16) we now have, for a real finite non-uniform stream, the relations corresponding to (1.4.3), (1.4.4) and (1.6.6) which were for an infinitesimal stream which could be assumed uniform. A point of special interest is that while the work relations (2.9.13) and (2.9.11) are identical in form to the infinitesimal case, the relation for energy transmission by heating is not exactly identical. It has the additional term $\sum_i M_i d(\tfrac{1}{2}v_{ri}^2)/M$ corresponding to changing kinetic energy of relative motion.

The presence of this additional term has an important significance in principle, which will now be discussed.

The argument throughout is based upon the assumption that no work extraction occurs from the non-uniformities. But the non-uniformity of motion is inevitably present in equation (2.9.14) since the kinetic energy of the relative motions is certainly part of the energy content of the substance. This is why the change in that quantity appears in the equation (2.9.16) for $\Delta\bar{Q}_t$. The correctness of its presence there will be readily appreciated by considering some simple cases.

(i) *Insulated rigid closed box.* Suppose we have suddenly enclosed a sample of substance in an insulated rigid box moving with $\bar{v}$. The process in the box is now $\Delta\bar{Q}_f = 0$, $\Delta\overline{W}_f = 0$ and $d\bar{V} = 0$. Hence also $dV^* = 0$ and (2.9.16) becomes

$$dU^* = -\sum_i M_i d(\tfrac{1}{2}v_{ri}^2)/M. \tag{2.9.17}$$

i.e. the kinetic energy of relative motion dies out against the internal mutual friction, and the internal energy increases accordingly.

(ii) *Adiabatic process.* In this case $\Delta\bar{Q}_t = 0$ and (2.9.16) becomes

$$\sum_i M_i d(\tfrac{1}{2}v_{ri}^2)/M = \Delta\overline{W}_f - (dU^* + p^*\,dV^*). \tag{2.9.18}$$

This corresponds to the real situation that the frictional work between the device and the substance is generating eddies with kinetic energy of relative motion, while simultaneously such eddies already present are dissipating their kinetic energy against mutual friction, and this appears as $dU^* + p^*\,dV^*$. The actual increment of relative kinetic energy is the net balance of the two effects.

(iii) *Impulse heating in a rigid closed stationary box.* In this case $\Delta\overline{W}_f = 0$ and $d\bar{V} = dV^* = 0$. Hence (2.9.16) becomes

$$\Delta\bar{Q}_i = dU^* + \sum_i M_i d(\tfrac{1}{2}v_{ri}^2)/M. \tag{2.9.19}$$

If we give a sudden impulse of energy by a quick local heating at a portion of the box, then during the interval when heat transmission is occurring we are changing internal energy non-uniformly, *and* causing convection currents which give the kinetic energy of relative motion. Once the heating is stopped, the situation becomes that of case (i), the convection currents die out and ultimately all the energy transferred in the impulse does go to increase of U^*.

Returning now to the general process equation we may write it in the form, analogous to (2.9.18),

$$\sum_i M_i d(\tfrac{1}{2}v_{ri}^2)/M = \Delta\bar{Q}_t + \Delta\overline{W}_f - (dU^* + p^* dV^*). \qquad (2.9.20)$$

This reveals that the actual increase of kinetic energy of relative motion can be seen as the combined generation effect of convection currents by heating and eddy creation by friction, less the dissipation of existing relative motion kinetic energy against mutual friction which appears as $dU^* + p^* dV^*$.

From the discussion of these examples it is evident that for a precise account of any real process in general, the term representing kinetic energy of relative motion must be present in the heat equation. Hence although the work relations (2.9.11) and (2.9.13) for the real case can be precisely identical in form to those for the infinitesimal uniform case, the heat relation must strictly be different.

In view of this we now set a limitation on the range of process which we consider. We shall specify that attention be restricted to processes in which the quantity $\sum_i M_i(\tfrac{1}{2}v_{ri}^2)/M$ and its change are negligible contributions to ϵ and to $d\epsilon$.

With this restriction (2.9.16) becomes

$$\Delta\bar{Q}_t = dU^* + p^* dV^* - \Delta\overline{W}_f. \qquad (2.9.21)$$

Alternatively this means that statement (2.9.21) is subject to an error of order of magnitude of the kinetic energy of relative motion actually present.

Within this possible error the relation for the finite non-uniform case is identical in form to the relation for the infinitesimal uniform case.

Equations (2.9.11), (2.9.13) and (2.9.21) are the basic thermodynamic relations for energy transmission by work and by heating as set down in Section 2.7 and used throughout the rest of the text. Their validity depends on the restriction to work transfer devices greater in size than the scale of non-uniformities, and they are subject to error of the order of magnitude of the kinetic energy of relative motions. Despite these restrictions they form the basis for applying the *first* law of thermodynamics to real processes.

The zeroth and second laws are directly applicable to real processes. Temperature is assumed to exist, and to be non-uniform in general in a real process. Energy transmission by heating occurs mutually between subordinate masses because of this non-uniformity in the direction determined by temperature difference and the second law. Similarly, energy transmission by heating occurs between the device and the substance in the direction determined by the second law and whatever temperature difference exists between the device and the substance.

3

THE CYCLIC PROCESS AND SOME IMPLICATIONS

If you have formed a circle to go into
Go into it yourself and see how you would do.

William Blake

3.1 Cyclic processes

We are now in a position to resume the development of thought in Chapter 1 on the possibility of obtaining work output from a process limited by ambient conditions. The conclusion was reached in Section 1.5 of that chapter that this possibility depends entirely upon whether we can vary the density of a substance in a cycle of operations so that, despite the presence of friction, the work output from the cycle is still positive. We have now proved that, provided we have the temperature difference necessary for the heating process to occur, we can have an energy transmission to a mass of substance such that, per unit mass,

$$\Delta Q_t = dU + p\,dV - \Delta W_f. \qquad (3.1.1)$$

Now let this occur through a cyclic process of the general kind described in Section 1.5. Then

$$Q_{t,c} = \oint \Delta Q_t = \oint dU + \oint p\,dV - \oint \Delta W_f \qquad (3.1.2)$$

$$= \oint dU + W_{t,c}. \qquad (3.1.3)$$

But from Chapter 2 (Section 2.6), the internal energy U is a property function of p and V, and hence its cyclic integral is zero.

$$\therefore \quad Q_{t,c} = W_{t,c}. \qquad (3.1.4)$$

Thus when a substance undergoes a complete cyclic process, the net energy transferred by heating to the reference substance is exactly equal to that transferred by work from it. This is true irrespective of how much or how little friction may occur in the cycle.

Equation (3.1.4) tells us that by a cyclic process the net energy transferred by one mode can be converted to net energy transferred by the other. It also makes a statement which is true whatever various detailed processes and devices may be involved at different parts of the cycle, and therefore permits us to proceed with a generality uncomplicated by concern for flow, non-flow, mass motion, kinetic energy, or potential energy. We shall of course at a later stage consider such matters and individual devices which may form part of a cycle.

The zeroth and second laws are also implicitly involved in this generalisation because it is these which govern in practice how the heating mode of transfer occurs. The full discussion of their effects on the conversion problem will, however, be delayed until later.

Meanwhile we continue this chapter with a section devoted to explaining just why the possibility of conversion between modes of energy transfer was so significant when first discovered. We do so not primarily because of the historical and sociological importance, since that is not the interest of this book, but because it illumines one of the major aspects of basic thermodynamics to which we have so far given little attention.

3.2 A matter of measurement and magnitude

Up to now we have discussed inter-relations between energy transmission by work and by heating without reference to measurement of either, although the possibility of measurement is everywhere implied. Before proceeding further we must look at the actual measurement principles which are available to us. One problem immediately arises – the measurement of energy transmitted by heating.

The measurement of energy transmitted by work is obviously simple in principle – since it is given by the definition as the product of force and distance. Thus when a force of one newton acts through a distance of one metre we say that the energy transmitted by work is one joule, and that one foot pound-force is transmitted when a one pound-force acts through a distance of one foot. But in the process of heating as we observe it – as for example in the stationary case of water being heated in a kettle by a flame, where from what we have found we believe that energy of amount $\int(dU+p\,dV-\Delta W_f)$ is transmitted – the only mechanically measureable portion is $\int(p\,dV-\Delta W_f)$. We can observe this in terms of our definition of energy by measuring the pressure and the volume change and the frictional force and displacement. But the only other observable is a rise of temperature – and while we may believe that this reflects the change in internal energy, we have no way – based on observations of the water alone – of determining what the amount is. Essentially what we do is to assume the first law, and rely upon experiments which are basically of the Joule type, to derive a correspondence between energy input by work measureable externally to the reference substance, and rise of temperature of the substance. Then when energy transmission by heating alone is found to give the same rise of temperature under the same conditions for the reference substance, we assume that the amount of energy transmitted is the same. From the second law, energy transmission by heating to some desired site will always be obtained provided we have some source at a higher temperature than that of the site.

It is within this context that we find just why the invention of the steam engine in the eighteenth century was of such immense sociological importance. One of the easiest practical ways of obtaining a source of energy transmission by heating – and the most readily available to early stages of technological development – is that of fire. The chemical reaction of combustible fuel with oxygen of the air produces, in ways for which the detail is unimportant at our present stage of study,† products at high temperature from which energy may be transmitted by heating. The amount of such energy can be determined by the principle described in the preceding paragraph. By such means it can be established that the amount of energy transmissible by heating from the combustion of only one pound mass of coal is of the order of 10^7 joules or about 7.4×10^6 ft lbf. It was this fact which, when the invention of the steam engine made possible the conversion of heating to work, revolutionised the world. Conversion of energy transmission from one mode to the other, although of scientific interest, would have been of far less significance if the order of magnitude of the energy released from combustion had been very much less. In fact, the combustion of one pound of the fuel can transmit by heating as much energy as could be transmitted by work from a ton of water falling, from the top of the highest mountain in Britain, to the sea. Even although only a part of this heating can be converted into work, it placed at man's disposal a controllable machine for doing work, in much higher quantities per unit of space occupied, than water wheels, windmills and treadmills.

This is the importance of the result established in Section 3.1. Power in useful amounts can be gained by heating arising from a source such as the combustion of fuel. Since we have to pay for and supply the fuel to where it is required, we naturally want to get as much work from the use

† These are discussed in Chapter 6.

of as little fuel as possible. Equation (3.1.4) shows that we get as work *all* the energy of the *net* transfer by heating. We therefore turn our attention now to this matter of *net* transfer. As a step in this study, we now examine in closer detail the nature of process thermodynamic path already introduced in Chapter 2 (Section 2.8).

3.3 Further discussion of the ideal frictionless process

This section amplifies the discussion of process and path given in Chapter 2 (Section 2.8).

Suppose that a substance has been made to undergo a process which takes it along path (a) from State 1 to State 2 in Figure 3.3.1. Now submit

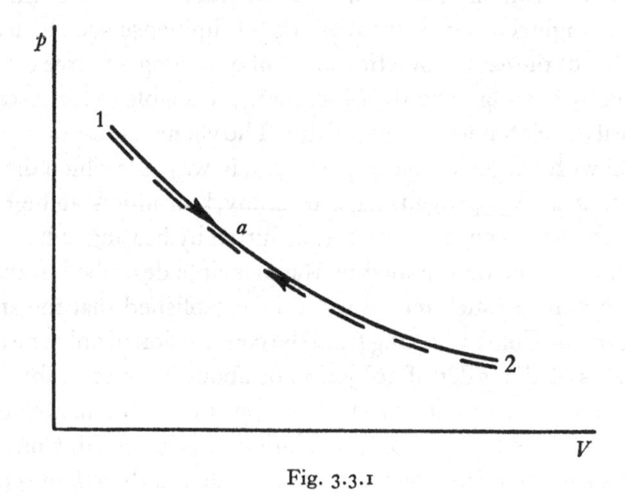

Fig. 3.3.1

it to a second process which takes it exactly back along the same path. (The return line is shown dotted just beside the original for illustration, but it should actually be superimposed.) Then we have

$$[a\int_1^2 (dU+pdV) = -[a\int_2^1 (dU+p\,dV). \qquad (3.3.1)$$

Hence

$$[a\int_1^2 (\Delta Q_t+\Delta W_f) = -[a\int_2^1 (\Delta Q_t+\Delta W_f). \qquad (3.3.2)$$

Now because of the nature of frictional dissipation we know that ΔW_f must always be positive. We have no other information about its possible value, and we cannot therefore in general go further from equation (3.3.2) except to say

$$[a\int_1^2 \Delta Q_t = -[a\int_2^1 \Delta Q_t - (\text{two positive terms}),$$

i.e.

$$[a\int_1^2 \Delta Q_t < -[a\int_2^1 \Delta Q_t, \qquad (3.3.3)$$

i.e. the energy transmitted by heating to the substance in the direct path must be less than the energy transmitted by heating from it in the return along the same path.† Hence although the return is along the same path, through the reverse sequence of thermodynamic states, the *process* is not a simple reversal, since more energy has to be rejected in the reverse than can be taken in the direct path.

But when there is no friction, F_f will always be zero and so $\int \Delta W_f$ is zero. In such an ideal case we then have

$$[a\int_1^2 \Delta Q_{t,0} = -[a\int_2^1 \Delta Q_{t,0}, \qquad (3.3.4)$$

where by $\Delta Q_{t,0}$ we mean the amount of energy taken in by heating under the ideal condition of zero friction. Thus with no friction, return along the same path would *exactly* reverse the whole process, rejecting the same amount of energy in the return as taken in the outward direction – provided also that thermal reversibility exists to enable the rejection to occur under the second law.

Again we note that the adjective 'reversible'‡ is properly applied to the *process*, *not* to the thermodynamic state path. We can in fact return the substance along exactly the same path, provided we suitably modify the amounts of energy rejection and intake by heating to compensate for frictional effects.

Thus any path on a p, V diagram is a real, possible, path for which we have the relations

$$[a\int_1^2 (dU + p\,dV) = [a\int_1^2 (\Delta Q_t + \Delta W_f) = [a\int_1^2 \Delta Q_{t,0},$$

$$\therefore \quad [a\int_1^2 \Delta Q_t = [a\int_1^2 \Delta Q_{t,0} - [a\int_1^2 \Delta W_f. \qquad (3.3.5)$$

† Although the matter is dealt with specifically in Section 3.4, we should nevertheless note meantime that these statements concerning energy transmission by heating *to* the substance along the *direct* path, and *from* the substance in the *reverse* path are algebraic and essentially *definitions* of direction. We have no way of knowing in general which direction implies a *positive* amount of energy transmission to the substance.

‡ The word 'reversible' is so well established in thermodynamic use that it must be retained with its current meaning. But we can of course – and do in practice – drive real, thermodynamically irreversible cycles in reverse. See Section 3.5 of this chapter.

From relation (3.3.5) we see that the energy which can be absorbed by a substance by heating in any thermodynamic path is always *less* in reality than the amount which could be absorbed by an ideal frictionless change along the same path. This requirement is algebraic, so that if the path is in fact one in which energy would be *rejected* in the ideal frictionless case, then in the real case *more* must be rejected to give the substance the same path of change.

These results can be expressed alternatively as follows.

(*a*) The maximum amount of energy which can be taken in by heating to a substance experiencing a definite thermodynamic change is that which is taken in when the change occurs by an ideal frictionless process.

(*b*) The minimum amount of energy which can be rejected by heating from a substance experiencing a definite thermodynamic change is that which is rejected when the change occurs by an ideal frictionless process.

3.4 Conversion properties in cyclic processess

From Section 3.1 of this chapter, we have seen that when a substance completes a closed cycle process, the net amount of energy taken in by heating or by working in the cycle is converted into energy given out by working or by heating, i.e.

$$\oint \Delta W_t = \oint \Delta Q_t. \tag{3.4.1}$$

From Section 3.3 we have seen that a given path has a maximum energy intake by heating and a minimum energy rejection by heating when it is frictionless. It follows therefore that the *net* amount of energy taken in by heating in a given cycle is a maximum when the cycle is performed by the ideal frictionless process. And of course the equal maximum amount of net work obtainable from a given thermodynamic cycle is that which would occur if there were no friction.

Now of course this statement in itself is obvious, since with $\Delta W_f = 0$ we see that the maximum value of $\oint \Delta Q_t = \oint \Delta W_t$ becomes $\oint p\,dV = -\oint V\,dp$. But there is another more important aspect which follows from the discussion in Section 3.3, which is not obvious. This concerns the *proportion* of energy *intake* by heating which is converted into work. First of all we must re-state, or recognise with emphasis, that all the *net* energy intake by heating is converted into energy output by work, whether or not friction is present. The situation is that the total energy intake is divided into two portions, i.e. the amount rejected by heating, and the amount converted to work. Since in practice we pay for the fuel which we use to provide the total energy intake by heating, and are usually concerned with obtaining work from it, we are naturally interested in the proportion in which the division between work and heating rejection occurs. The conclusions in Section 3.3 show that for a given path in which energy is taken in by heating, the amount taken in will be a maximum when there is no friction – and in a path in which energy is given out by heating, the amount given out will be a minimum when there is no friction. Thus let the cycle of Figure 3.4.1 be distinguished by a portion from *a* to *b* via *s* as a path in which energy Q_s is supplied by heating, and the portion from *b* to *a* via *r* as one in which energy Q_r is rejected, i.e. Q_r is a positive quantity, defined as

$$-[r]\int_b^a \Delta Q_t.$$

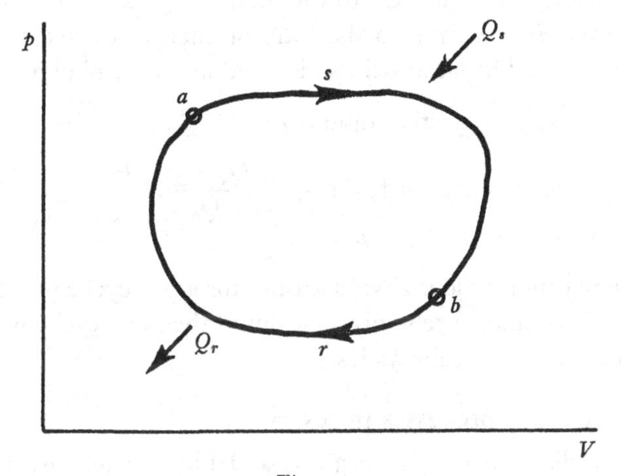

Fig. 3.4.1

Then
$$W_{t,c} = \oint \Delta W_t = \oint \Delta Q_t = Q_s - Q_r.$$

If the same cycle were performed in an ideal frictionless process we should have similarly
$$W_{t,c,0} = Q_{s,0} - Q_{r,0}.$$

We already know that $W_{t,c} < W_{t,c,0}$. But the important additional result concerning the proportions is that since

$$Q_{r,0} < Q_r \quad \text{(from conclusion (}b\text{) at the end of Section 3.3),}$$

and $Q_{s,0} > Q_s$ (from conclusion (*a*) at the end of Section 3.3),

$$\therefore \quad \frac{Q_{r,0}}{Q_{s,0}} < \frac{Q_r}{Q_s}. \tag{3.4.2}$$

$$\therefore \quad \frac{Q_{s,0}-Q_{r,0}}{Q_{s,0}} > \frac{Q_s-Q_r}{Q_s},$$

i.e.

$$\frac{W_{t,c,0}}{Q_{s,0}} > \frac{W_{t,c}}{Q_s}. \tag{3.4.3}$$

Thus not only is the work performed and the total energy intake by heating in a real cycle with friction less than in the ideal frictionless case, but the *proportion* of the total energy intake converted into work is also less.

We shall give the name *conversion ratio* to the ratio of energy transmission by work from the cycle to the total energy supply by heating, and the name *rejection ratio* to the ratio of energy rejected to energy supply by heating. The latter will be denoted by the symbol ρ.

Thus we have

$$\text{Rejection ratio} = \rho = Q_r/Q_s. \tag{3.4.4}$$

$$\text{Conversion ratio}\dagger = 1-\rho = \frac{Q_s-Q_r}{Q_s} = \frac{W_{t,c}}{Q_s}. \tag{3.4.5}$$

Then always

$$\rho > \rho_0. \tag{3.4.6}$$

Thus the rejection ratio is always greater for a real cyclic process than for the ideal frictionless cyclic process, round the same cyclic path, and so the conversion ratio is always less.

3.5 Driving cyclic processes in reverse

Consider the diagram in Figure 3.4.1, and imagine that we transmit energy by work so as to make the substance experience the same cycle in reverse, i.e. it goes from *b* to *a* via *s* and from *a* to *b* via *r*. We put in a net amount of energy $W_{t,c}^*$ by work. All the arrows are reversed, and Q_r^* is *absorbed* along the lower curve while Q_s^* is *rejected* along the upper curve. The relation

$$W_{t,c}^* = Q_s^* - Q_r^* \tag{3.5.1}$$

† It will be seen that by conversion ratio we mean what has usually been termed the 'thermodynamic efficiency', or merely the 'efficiency'. The name is somewhat unfortunate and misleading, since no actual loss of *energy* has occurred. What is really intended is *conversion efficiency*, and it seems preferable to emphasise that the matter is one of conversion. The essential corollary that there is also energy rejection, is more definitely implied by the term conversion ratio, as well as by the deliberate use of rejection ratio. Hence we do not suggest a separate symbol for conversion ratio, preferring to write it as $(1-\rho)$.

is still true. The values of $W_{t,c}^*$, Q_s^*, and Q_r^* are not of course the same as $W_{t,c}$, Q_s, and Q_r for the same direct cycle of Figure 3.4.1, if friction is present. But they are identical for the ideal frictionless – thermodynamically reversible process.

This reverse driving of a real irreversible, practical, thermodynamic cycle is quite ordinary and common. The fact that the *process* cannot be actually reversed in all its details is what we mean by thermodynamic irreversibility as discussed in previous sections, but this does not exclude the possibility of reverse driving. But what utility does this procedure have? The conversion of energy available from heating into work had the obvious advantage and utility discussed in Section 3.2 of this chapter. But what useful thing can we do if we drive the engine round the thermodynamic cycle in the opposite way? At first sight this reverse driving appears merely to convert the energy transmission by work into energy transmission by heating, since the amount of energy rejected Q_s^*, is of course $Q_r^* + W_{t,c}^*$.

But we know that such conversion can be done very simply merely by a force working against friction, without requiring the elaborate devices needed to make a substance go through a thermodynamic cycle. Hence this reverse driving of a cycle cannot be of much utility unless it does something much more than mere conversion. And of course it does do something much more significant. To understand this we must turn our attention to the zeroth and second laws. These were mentioned at the end of Section 3.1 of this chapter as being relevant, but so far our discussion of cyclic processes has *used* only the first law. We now proceed to use the others.

For convenience in reference let us for the present adopt the terminology *engine* for the direct cycle operating so that Q_s is absorbed by heating and Q_r rejected, with the balance appearing as work output $W_{t,c}$; and *reversed engine* for the reverse driven cycle. In the latter case Q_r^* is absorbed by heating, a work input $W_{t,c}^*$ is given, the sum appearing as Q_s^* rejected.

Now consider first the engine case and apply the zeroth and second laws to it. The substance will have temperature, which will in general vary around the cycle. The argument is simplified if we direct attention to a particular cycle where we assume that the energy supply by heating, Q_s, and the rejection, Q_r, occur to and from the substance respectively at constant temperature. The substance temperature may be changed from θ_s to θ_r, and back again by two processes defined to be such that no energy is transmitted to or from the substance by heating. We illustrate the cycle so defined by the diagram Figure 3.5.1.†

From the second law, to supply Q_s to the substance by heating, there has to be some source at a temperature θ_{ss} higher than θ_s. Similarly, to reject Q_r there has to be some sink at a temperature θ_{rr} lower than θ_r. $\theta_{ss}-\theta_s$ and $\theta_r-\theta_{rr}$ need only be infinitesimal in principle. But we can go further with the second law to establish a relation between θ_s and θ_r.

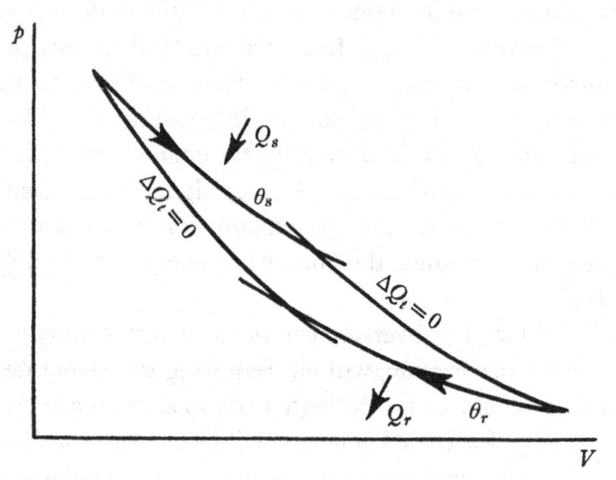

Fig. 3.5.1

At first we have two alternative assumptions, i.e.

$$\text{(i) } \theta_s > \theta_r, \quad \text{and} \quad \text{(ii) } \theta_r > \theta_s.$$

Now if $\theta_s > \theta_r$, we should have $\theta_{ss} > \theta_{rr}$. Thus observing the overall results of this process we should see energy transmission by heating occurring from higher to lower temperature, with energy transmission by work appearing as an output by-product. If, however, $\theta_r > \theta_s$, it would be possible to have $\theta_{rr} > \theta_{ss}$, and so on assumption (ii) we should observe the overall result as energy transmission by heating occurring from lower to higher temperature, again with energy transmission by work as an output by-product. Now if positive energy output by work, however small, can be obtained, we presume that the process can and will happen spontaneously. Thus we should observe overall transmission by heating

† The diagram happens to show positive energy intake by heating at the upper curve with expansion. This is irrelevant. For some substances at some conditions energy intake by heating will occur with compression as along the lower curve. Note carefully that the argument following does not depend on the relative positions in the diagram. See also note on page 63.

from the lower temperature θ_{ss} to the higher temperature θ_{rr}, under assumption (ii), and from the higher temperature θ_{ss} to the lower temperature θ_{rr} under assumption (i). The second law therefore implies that assumption (ii) is unacceptable and that assumption (i) is acceptable. We conclude, therefore, that to have an engine operating between two constant temperatures with production of net output energy by work requires essentially that θ_s, the temperature of energy supply by heating, must be greater than θ_r, the temperature of energy rejection by heating.†

Secondly, consider the reversed driving of the cycle in Figure 3.5.1. Again we consider the corresponding alternative assumptions (i) $\theta_r^* > \theta_s^*$, and (ii) $\theta_s^* > \theta_r^*$. Since Q_r^* is absorbed this time at θ_r^*, there must be a source at $\theta_{rr}^* > \theta_r^*$ and to reject Q_s^* at θ_s^* there has to be a sink at $\theta_{ss}^* < \theta_s^*$.

With assumption (i) we have energy transmission by heating entirely from higher to lower temperatures, together with the input of energy transmission by work. This is entirely permitted and possible by the second law.

Now, considering assumption (ii), it would be possible to have $\theta_{ss}^* > \theta_{rr}^*$. Hence the overall observation would be energy transmission from θ_{rr}^* to θ_{ss}^*, i.e. from a lower to a higher temperature, with energy transmission by work as an *input*. Thus the situation under assumption (ii) could not occur and would not be observed spontaneously.

This statement of negation is consistent with the second law. The affirmation, that the process can occur with $\theta_r^* < \theta_s^*$ provided there is work input, cannot positively be said to be proved by the second law, but is not denied by it.

We can present the results of this analysis as a table.

Assumption	Engine (work output)	Reversed engine (work input)
Energy intake by heating at higher temperature, rejection at lower	Permitted *A*	Permitted *B*
Energy intake by heating at lower temperature, rejection at higher	Denied *C*	Permitted *D*

The various conclusions are indicated as *A*, *B*, *C*, *D*. The value and utility of conclusion *A* has already been discussed in Section 3.2, and is

† In the diagram the temperature θ_s now proved to be the greater of the two, is shown as the upper curve. This is irrelevant. Some substances at some conditions expand with falling temperature. Note carefully that the argument given does not depend on the relative positions in the diagram. See also note on page 62.

undoubted. Conclusion C is a fact, the obverse correlate of A. Conclusion B, in face of A, indicates a possible, but pointless process. By taking in energy by heating at a higher temperature and rejecting at a lower we can obtain work output. To do the same, but give work input, is of little value. It implies that the work input is entirely dissipated against friction. Conclusion D, however, means that by driving the cycle in reverse we *can* absorb energy by heating from a lower temperature and reject at a higher. The utility of this is not in the conversion, but in the implication that we can for example, take energy by heating from water at 0° C and reject it to the ambient atmosphere at 15° C. Hence we can freeze the water, and cool matter to temperatures lower than that of the environment, provided we use a work input.

Thus the answer to our query about the utility of the reverse driven cycle is that it gives the possibility of rendering and keeping something below the environment temperature. Just as is the production of output work permitted under conclusion A, this process, under D is an essential feature of modern civilisation. We may term it in general *refrigeration*, and a reversed engine operating under the conditions of D will now be termed a *refrigerator*.

The argument leading to the possibility of a refrigerator has been based for convenience on the reverse driving of a particular engine (Figure 3.5.1) absorbing and rejecting energy by heating at two constant temperatures, the temperature changes occurring without energy exchange by heating. This establishes a significance and meaning for the reversed driving of any cycle, since the same qualitative characteristics will apply, although complicated by detail of temperature variation. Nevertheless just as in Section 3.4 we established a criterion, the rejection ratio, to characterise an engine, we can now proceed similarly to a criterion to characterise a refrigerator.

3.6 Performance of a refrigerator

In the engine case, our interest, for utility and economic reasons ultimately, is in the conversion ratio $W_{t,c}/Q_s$, i.e. $1-\rho$. In the refrigerator case, however, the interest is in knowing how much energy we can transfer from lower to higher temperature per unit of input work. This ratio is $Q_r^*/W_{t,c}^*$ and is usually called the performance coefficient or performance factor of the refrigerator, and will be defined by

$$K^* = \frac{Q_r^*}{W_{t,c}^*}. \tag{3.6.1}$$

Since this time Q_s^* is being rejected, we know that for the real case, Q_s^* must be greater than $Q_{s,0}^*$ for a frictionless case, and also that $Q_r^* < Q_{r,0}^*$.

Hence

$$Q_s^* - Q_r^* > Q_{s,0}^* - Q_{r,0}^*.$$

$$\therefore \quad W_{t,c}^* > W_{t,c,0}^*, \tag{3.6.2}$$

and since

$$Q_r^* < Q_{r,0}^*, \tag{3.6.3}$$

$$\frac{Q_r^*}{W_{t,c}^*} < \frac{Q_{r,0}^*}{W_{t,c,0}^*}. \tag{3.6.4}$$

Thus the performance coefficient for the real frictional system is necessarily less than that of the ideal frictionless cycle, i.e. we get less refrigeration effect from the input of a given amount of energy by work.

Since the frictionless case is thermodynamically reversible we know that $Q_{r,0}^*$, $Q_{s,0}^*$, and $W_{t,c,0}^*$ are respectively equal to $Q_{r,0}$, $Q_{s,0}$, and $W_{t,c,0}$ for the direct engine cycle.

Hence

$$K_0^* = \frac{Q_{r,0}^*}{W_{t,c,0}^*} = \frac{Q_{r,0}^*}{Q_{s,0}^* - Q_{r,0}^*} = \frac{\rho_0}{1-\rho_0}. \tag{3.6.5}$$

For the frictional case we cannot assume that the dissipation and energy intake and rejection by heating are the same when driven in reverse as in the direct system and we cannot obtain an equation between K^* and ρ to correspond to equation (3.6.5). But from (3.6.4) and (3.6.5) we have

$$K^* < \frac{\rho_0}{1-\rho_0}, \tag{3.6.6}$$

and since $\rho > \rho_0$ we have also

$$K^* < \frac{\rho}{1-\rho}. \tag{3.6.7}$$

Thus, for example, an ideal frictionless reversible engine cycle with a rejection ratio of 0.6 and a conversion ratio of 0.4 will have a performance factor of 3/2, but a real cycle with a rejection ratio of 0.7 and conversion ratio of 0.3 will have a performance factor $< \frac{7}{3}$ and we cannot be more definite.

An important point is contained here. It should be noted that for the ideal reversible cycle, a high conversion ratio ('efficiency'), i.e. large $1-\rho_0$ and small ρ_0 means a low performance factor and vice versa. Thus to get most economic refrigeration we should desirably produce output work from a reversible engine with high conversion ratio, and use it to drive in reverse a different reversible engine with a low conversion ratio.

Denoting these two cycles respectively by suffixes 1 and 2, the arrangement will be that $Q_{s,1,0}$ is taken in by heating and $Q_{r,1,0}$ rejected, and the work $Q_{s,1,0}-Q_{r,1,0}$ is used to drive the other in reverse so as to absorb $Q^*_{r,2,0}$ and reject $Q^*_{s,2,0}$.

Since the net energy transmission by work is equal in each cycle, we have

$$Q_{s,1,0}-Q_{r,1,0}=Q^*_{s,2,0}-Q^*_{r,2,0}, \tag{3.6.8}$$

i.e. the net energy transmission by heating is also equal in each cycle. We have also

$$Q_{s,1,0}-Q^*_{s,2,0}=Q_{r,1,0}-Q^*_{r,2,0}. \tag{3.6.9}$$

From the temperature relations established in Section 3.5 we know that $Q_{s,1,0}$ is energy taken in by heating at temperature levels higher than those of $Q_{r,1,0}$ and that $Q^*_{s,2,0}$ is energy rejected by heating at temperature levels higher than those of $Q^*_{r,2,0}$. Thus $Q_{s,1,0}-Q^*_{s,2,0}$ is the net energy taken in by heating at the higher ranges of temperature and the equivalent $Q_{r,1,0}-Q^*_{r,2,0}$ is the net energy rejected by heating at the lower ranges of temperature. We have not, however, so far defined any relative position for the temperatures in the respective cycle. The discussion is made somewhat difficult by the variation of temperature which in general will occur around a cycle such as in Figure 3.4.1, and to reduce this difficulty we resort again to a particular cycle similar to that of Figure 3.5.1 where θ_s upper and θ_r lower constant temperature processes apply.

3.7 The Carnot cycle

Consider Figure 3.7.1, which shows essentially the same cycle as that of Figure 3.5.1.

As compared with Section 3.5, however, we now make the additional specification that at every point in the cycle the process is ideally frictionless and thermodynamically reversible. Thus for the engine case $Q_{s,0}$ is absorbed at θ_s and $Q_{r,0}$ rejected at θ_r, and in reverse drive $Q^*_{r,0}$ is absorbed at θ_r and $\theta^*_{s,0}$ rejected at θ_s. The higher of the two temperatures is θ_s.

The processes *bc* and *da* change temperature with no absorption or rejection of energy by heating. This cycle, when frictionless and thermodynamically reversible, is called the Carnot cycle, after the French engineer, Sadi Carnot, who first described its essentials.

We remind the reader at this stage that the numeration of θ_s and θ_r is still, at the present stage of our argument, on some quite arbitrary basis. What we can say for certain is that $\theta_s > \theta_r$.

It is also usefully recalled here that the actual amounts of energy transmission can each be made in proportion, as large or as small as we please, by using different masses of substance in the cycle. Also the chemical nature of the substance in the cycle may be anything we please.

Now let us consider two such Carnot cycle processes C_1 and C_2 operating with different masses m_1 and m_2 of different substances, but operating between the same upper and lower temperatures θ_s and θ_r. Let C_1 operate as an engine so as to drive C_2 in reverse as a refrigerator, and let its energy output by work be just sufficient to do this.

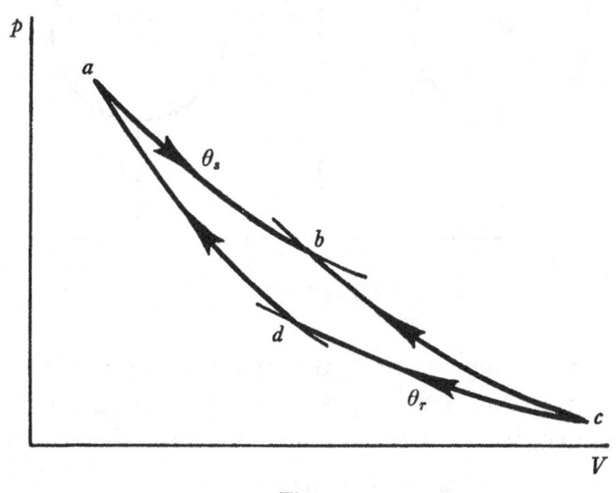

Fig. 3.7.1

Then the circumstances may be represented by the diagram of Figure 3.7.2.

In C_1 energy of amount $m_1 Q_{s,1,0}$ is taken in by heating at θ_s and $m_1 Q_{r,1,0}$ is rejected at θ_r, producing energy output by work $\mathscr{W}$ such that

$$\mathscr{W} = m_1(Q_{s,1,0} - Q_{r,1,0}). \tag{3.7.1}$$

In C_2 energy of amount $m_2 Q^*_{r,2,0}$ is taken in by heating at θ_r and $m_2 Q^*_{s,2,0}$ is rejected at θ_s, as a result of energy input by the work $\mathscr{W}$, so that

$$m_2(Q^*_{s,2,0} - Q^*_{r,2,0}) = \mathscr{W}. \tag{3.7.2}$$

From (3.7.1) and (3.7.2) we have

$$m_1 Q_{s,1,0} - m_2 Q^*_{s,2,0} = m_1 Q_{r,1,0} - m_2 Q^*_{r,2,0}, \tag{3.7.3}$$

i.e. the net amount of energy *taken* by heating at θ_s is equal to the net amount of energy *rejected* by heating at θ_r.

Since all energy transmitted by work is completely produced and absorbed within the joint system, the only result which an external observer will notice is the transmission of energy by heating from a

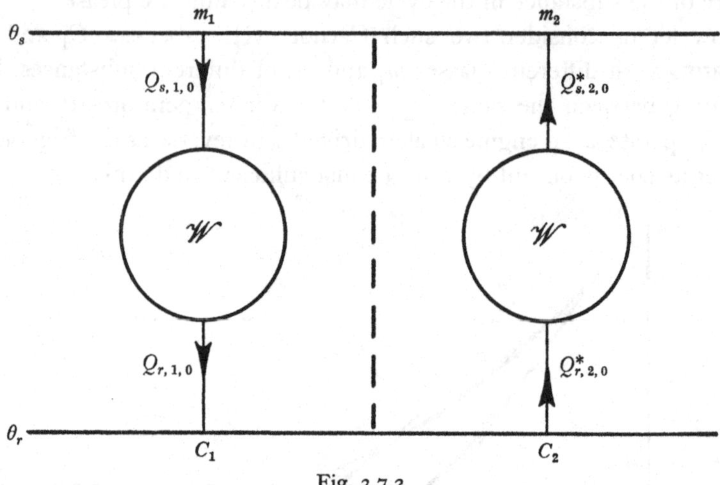

Fig. 3.7.2

temperature θ_s to a temperature θ_r, the amount being that on each side of equation (3.7.3). Since $\theta_s > \theta_r$ the energy transmission by heating *cannot* be negative, from the second law. It follows therefore that

$$m_1 Q_{s,1,0} \geqslant m_2 Q^*_{s,2,0},$$

$$\therefore \quad \frac{m_1 Q_{s,1,0}}{m_2 Q^*_{s,2,0}} \geqslant 1. \tag{3.7.4}$$

Also from equation (3.7.3) and since

$$Q_{r,1,0} = \rho_{1,0} Q_{s,1,0},$$

and

$$Q^*_{r,2,0} = \rho_{2,0} Q^*_{s,2,0},$$

$$m_1 Q_{s,1,0}(1-\rho_{1,0}) = m_2 Q^*_{s,2,0}(1-\rho_{2,0}). \tag{3.7.5}$$

Substituting in (3.6.8) we have therefore

$$\frac{1-\rho_{2,0}}{1-\rho_{1,0}} \geqslant 1. \tag{3.7.6}$$

Relation (3.7.6) is very interesting. It shows that, to meet the requirements of the second law that energy transmission by heating must go from the higher to the lower temperature, the amounts of energy are irrelevant, the masses are irrelevant, and the only condition is that the respective rejection or conversion ratios should satisfy (3.7.6). But of course the suffixes 1 and 2 have no absolute meaning – they distinguish only the driving cycle from the driven cycle. If they were connected the opposite way, so that C_2 with m_2 drove C_1 with m_1 we should find similarly the result

$$\frac{1-\rho_{1,0}}{1-\rho_{2,0}} \geqslant 1. \tag{3.7.7}$$

The only way to satisfy both (3.7.6) and (3.7.7) is to have

$$\frac{1-\rho_{1,0}}{1-\rho_{2,0}} = 1,$$

$$\therefore \quad \rho_{1,0} = \rho_{2,0}. \tag{3.7.8}$$

Equation (3.7.8) is of fundamental importance. We have already seen that the rejection ratio for any cyclic process is a minimum when it is frictionless. Since we choose *any* two Carnot cycles using any substances and amounts we now see that this minimum value is identical for *all* Carnot cycles, whatever the quantities of mass and energy, operating between the same two temperatures. Hence the rejection ratio of a Carnot cycle cannot be a function of substance, energy quantities, nor of anything else, except the temperatures between which it operates.

3.8 A basic concept of temperature

The conclusion of the preceding section was that the rejection ratio of an ideal Carnot cycle could depend on nothing other than the two temperatures between which it operates. This means that for a Carnot cycle operating between temperatures θ_a and θ_b we should properly say that the rejection ratio ρ must be regarded as $\rho_{a,b}$ and be expressible as some function of θ_a and θ_b, i.e.

$$\rho_{a,b} = f(\theta_a, \theta_b). \tag{3.8.1}$$

But consider now Figure 3.8.1.

We see that a Carnot cycle between θ_1 and θ_n can be considered as any two cycles such as that between θ_1, θ_2 and θ_2, θ_n and so on. In general consider the cycle θ_1, θ_n to be composed of θ_1, θ_i, and θ_i, θ_n where θ_i is a temperature intermediate between θ_1 and θ_n. Energy rejected by heating at θ_i in the θ_1, θ_i cycle is energy absorbed by heating at θ_i for the θ_i, θ_n cycle.

Thus
$$\rho_{1,i} = \frac{Q_{i,0}}{Q_{1,0}}; \tag{3.8.2}$$

$$\rho_{i,n} = \frac{Q_{n,0}}{Q_{i,0}}. \tag{3.8.3}$$

Multiplying (3.8.2) and (3.8.3) we obtain

$$\rho_{1,i} \cdot \rho_{i,n} = \frac{Q_{n,0}}{Q_{1,0}} = \rho_{1,n}. \tag{3.8.4}$$

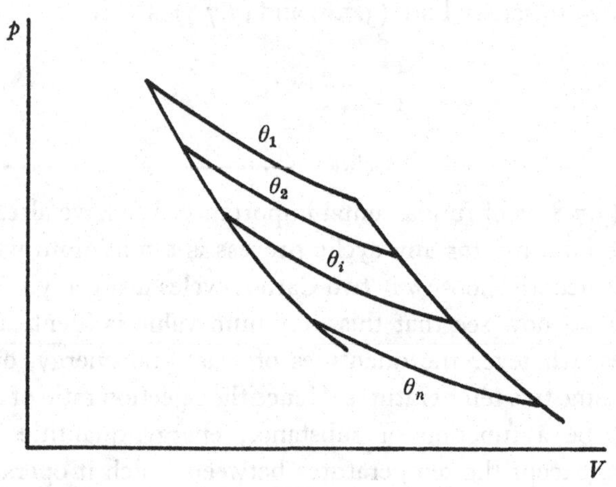

Fig. 3.8.1

Hence we establish a condition which must be satisfied by the function $f(\theta_a, \theta_b)$. This is

$$f(\theta_1, \theta_i) \cdot f(\theta_i, \theta_n) = f(\theta_1, \theta_n). \tag{3.8.5}$$

The condition (3.8.5) can always be satisfied if $f(\theta_a, \theta_b)$ is of the form

$$f(\theta_a, \theta_b) = \frac{T(\theta_a)}{T(\theta_b)},$$

or
$$f(\theta_a, \theta_b) = \frac{T(\theta_b)}{T(\theta_a)} \tag{3.8.6}$$

where $T(\theta)$ is *any* monotonic increasing function of θ. The two alternatives are merely reciprocals of each other, and either form may be used. We adopt the form

$$\rho_{a,b} = \frac{Q_b}{Q_a} = \frac{T(\theta_b)}{T(\theta_a)}. \tag{3.8.7}$$

The element of logic in our choice of this form rather than the reciprocal is that the rejection ratio is the ratio of an occurrence quantity at θ_b to another at θ_a, so that it seems in keeping to take the ratio of the value of the function T at θ_b to that at θ_a.

Following this procedure further we can now imagine the series of Carnot cycles in Figure 3.8.1 being such that each transmits an identical net amount of energy by work – so that

$$Q_1 - Q_2 = Q_2 - Q_3 = \ldots = Q_i - Q_{i+1}.$$

$$\therefore \quad Q_1\left[\frac{T(\theta_1)-T(\theta_2)}{T(\theta_1)}\right] = Q_2\left[\frac{T(\theta_2)-T(\theta_3)}{T(\theta_2)}\right] = \ldots \text{ etc.} \tag{3.8.8}$$

$$\therefore \quad T(\theta_1)-T(\theta_2) = [T(\theta_2)-T(\theta_3)]\frac{Q_2}{Q_1}\cdot\frac{T(\theta_1)}{T(\theta_2)} \tag{3.8.9}$$

$$= T(\theta_2)-T(\theta_3) = T(\theta_3)-T(\theta_4) = \ldots \text{etc.}$$

Thus the intervals $T(\theta_i)-T(\theta_{i+1})$ are equal for each cycle. In other words, equal intervals of the function $T(\theta)$ occur when the ideal engine does equal amounts of work between successive values of temperature θ.

Now once again we recall that throughout our discussion so far the notion of temperature θ, although defined conceptually quite clearly in the zeroth law, has been arbitrary and somewhat vague in its numeration. We have had in mind scales of measurement based on particular devices such as fluid expansion thermometers, electrical resistance thermometers, etc., i.e. based on temperature dependent properties of particular substances. The result just established means that, since temperature is the *sole* factor on which the rejection ratio of an ideal reversible Carnot cycle depends, then the rejection ratio of such a cycle provides a temperature dependent property which is independent of chemical constitution, pressure, or anything else. Hence it can be used as a measure of temperature in a sense which is quite absolute. For since the function $T(\theta)$ is an *arbitrary* monotonic increasing function of θ, then we may equally regard θ as an arbitrary monotonic increasing function of T, so that $T(\theta)$ becomes the measure of temperature – or simply *the* temperature value.

Thus we re-write equation (3.8.7)

$$\rho_{a,b} = \frac{T_b}{T_a} = \frac{Q_b}{Q_a}, \tag{3.8.10}$$

and the equal temperature intervals between which equal amounts of work are done by the ideal Carnot engine are

$$T_1 - T_2 = T_2 - T_3 = \ldots \text{etc.} \tag{3.8.11}$$

The measurement may be further specified by defining $T = 0$ as the temperature at which the energy rejection by heating from a substance undergoing an ideal Carnot cycle is zero.

The temperature established in this way is called the *thermodynamic temperature.*

Thus an ideal Carnot engine taking in energy by heating at a thermodynamic temperature T would give complete conversion to work, if the lower temperature of the cycle were zero. And equal amounts of work done in such a cycle between temperature intervals ΔT_1 and ΔT_2 will define these intervals as equal. This gives a clear and absolute significance to the concept of zero temperature. It is absolute in a sense which is quite absent from the arbitrary definitions of zero on, for example, the Celsius or Fahrenheit scales.

However the *enumeration* of any temperature other than zero will depend entirely on the size of unit division. Thus we now have a situation exactly analogous to the measurement of mass. We know what is meant by zero mass, i.e. that has an absolute significance, but whether we give the number 1, or 0.453 or 16 to a particular actual mass depends upon whether our division unit is the pound, the kilogram, or the ounce, which are all arbitrary.

Similarly, the numerical value of a thermodynamic temperature, although the concept itself is absolute, will depend on what arbitrary unit interval we adopt. We shall come later to the conventional choice of intervals, but meantime there is an aspect of prior concern to be discussed.

This is the question of actual thermometry. While the foregoing discussion has given a clear *significance* to temperature, the device of a Carnot cycle engine is impossible to realise – and even if it were not so, it would still be a very clumsy kind of thermometer. The problem may be realised as follows. Granted that we may choose an arbitrary division unit, how are we to know what number of such units to assign to a given temperature to obtain the thermodynamic value? Suppose for example we choose the Celsius unit division, i.e. one hundredth of the temperature interval between the freezing point and the boiling point of pure water. How do we determine – in a reasonably practical way – the number of such units between the absolute zero and the freezing point or the boiling point? Chapter 4 provides the groundwork of thermodynamic logic which ultimately enables this question to be answered with confidence.

4

PROPERTIES OF SUBSTANCE AND THEIR INTER-RELATIONS

Theory as such is of no use except in so far as it makes us believe in the coherence of phenomena.

Goethe

4.1 Entropy

In this section, all reference to temperature, unless otherwise mentioned, will mean thermodynamic temperature.

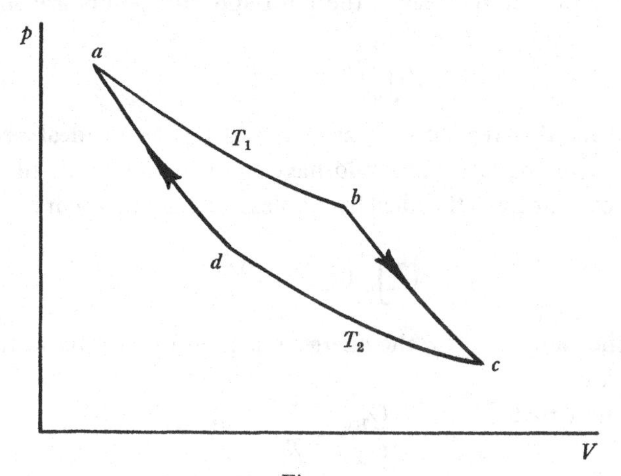

Fig. 4.1.1

Consider Figure 4.1.1, which shows the p, V diagram for a substance being subjected to the cyclic process consisting of the following steps, all of which may be regarded as real, i.e. irreversible, with friction present.

Step. 1. Expansion from a to b at constant temperature T_1.

Step 3. Compression from c to d at constant temperature T_2.

The connecting steps from b to c and from d to a will each be done by processes such that for the substance we have

$$dU + p\,dV = 0, \tag{4.1.1}$$

i.e. (4.1.1) is essentially a differential equation for the paths bc and da.

Since we have not limited the case to reversible processes, we note that, in general, since

$$\Delta Q_t = dU + p\,dV - \Delta W_f,$$

energy must be rejected from the substance by heating along paths *bc* and *da*, the positive amounts rejected being

$$\int_b^c \Delta W_f \quad \text{and} \quad \int_d^a \Delta W_f,$$

while from *a* to *b* and from *c* to *d* the energy absorption and rejection will include effects of friction in addition to the amounts arising from $dU + p\,dV$. But let us concentrate on $dU + p\,dV$ only.

We have noted in Chapter 2 (Section 2.6) that the value of

$$[T_1] \int_a^b (dU + p\,dV)$$

is completely determined, since the path and end points are specified. Hence

$$[T_1] \int_a^b (dU + p\,dV)$$

is the same for this real process as it would be for an ideal reversible process. But for the latter it would have the value $Q_{1,0}$, i.e. the energy absorption by heating in the ideal frictionless case. The similar conclusion is true for

$$-[T_2] \int_c^d (dU + p\,dV),$$

i.e. it has the value of $Q_{2,0}$, the energy rejection by heating in the ideal frictionless case.

But we know that

$$\frac{Q_{1,0}}{Q_{2,0}} = \frac{T_1}{T_2}. \tag{4.1.2}$$

$$\therefore \quad \frac{[T_1] \int_a^b (dU + p\,dV)}{-[T_2] \int_c^d (dU + p\,dV)} = \frac{T_1}{T_2}. \tag{4.1.3}$$

$$\therefore \quad [T_1] \int_a^b \frac{(dU + p\,dV)}{T_1} + [T_2] \int_c^d \frac{(dU + p\,dV)}{T_2} = 0. \tag{4.1.4}$$

But from (4.1.1) dividing by the general temperature T as a variable in the path, and integrating, we have

$$\int_b^c \frac{(dU + p\,dV)}{T} + \int_d^a \frac{(dU + p\,dV)}{T} = 0. \tag{4.1.5}$$

Adding (4.1.4) and (4.1.5) gives

$$\oint \frac{(dU+p\,dV)}{T} = 0. \tag{4.1.6}$$

Equation (4.1.6) shows that the cyclic integral of the quantity $(dU+p\,dV)/T$ is zero.

Now admittedly we have defined a particular sequence of processes. But if we consider a point X such as shown on Figure 4.1.2, we see that an infinite number of cycles of the same specification can be drawn to

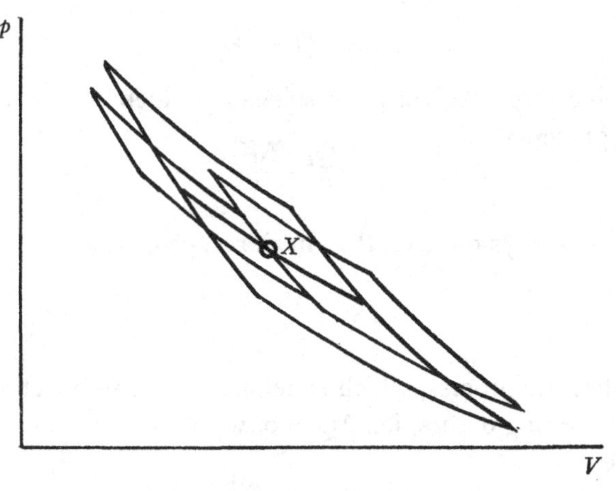

Fig. 4.1.2

include this point. A few such are shown in the diagram of Figure 4.1.2. It follows that by any such cyclic process beginning at X and finishing at X we will have

$$\oint \frac{(dU+p\,dV)}{T} = 0.$$

This fact implies that there is some state *property* of the substance, which we shall denote by S, such that

$$dS = \frac{dU+p\,dV}{T}. \tag{4.1.7}$$

Only if such a property exists can a cyclic process have a zero integral, since this means that we return to the same initial value of all state properties.

The property S, which is defined by equation (4.1.7), is called the *entropy*, and is of considerable importance in the further developments of thermodynamics. Meantime we note certain particular relevant points.

First of all we can now give a name to the type of change defined by equation (4.1.1). It is clear that in such a change the entropy of the reference substance remains constant. Thus such a process path is a *constant entropy* path or an *isentropic* path.

Secondly we can now see the thermal first law in a rather different light. We recall that the energy transmission by heating is

$$\Delta Q_t = dU + p\,dV - \Delta W_f. \tag{4.1.8}$$

Hence we now see
$$\Delta Q_t = T\,dS - \Delta W_f, \tag{4.1.9}$$

or
$$T\,dS = \Delta Q_t + \Delta W_f. \tag{4.1.10}$$

Thus the increase of entropy of a substance which undergoes *any* real irreversible process is

$$dS = \frac{\Delta Q_t}{T} + \frac{\Delta W_f}{T}. \tag{4.1.11}$$

Since ΔW_f is always positive, then in all real processes

$$dS > \frac{\Delta Q_t}{T}. \tag{4.1.12}$$

For an adiabatic process, which is defined as one in which no energy exchange by heating occurs, i.e. $\Delta Q = 0$, we always have an increase of entropy given by

$$dS_{\text{adiabatic}} = \frac{\Delta W_f}{T}. \tag{4.1.13}$$

Only in the ideal frictionless case can we have a process which is *both* adiabatic and isentropic. In any real process, to maintain the reference substance at constant entropy requires the extraction of energy by heating, i.e. we must have negative ΔQ_t. Similarly we can reduce the entropy of a reference substance by extraction of sufficient energy by heating to ensure that $dU + p\,dV = T\,dS$ for the substance is negative.

And here we are led to realise a further basic significance about the property entropy. From the second law if we exchange energy by heating from a reference substance A to another B we know that if

$$\Delta Q_t' = m_A(dU_A + p_A\,dV_A)$$

is to be negative and the equal quantity

$$\Delta Q_t' = m_B(dU_B + p_B\,dV_B)$$

is to be positive, then for the transfer by heating to take place we must have $T_B < T_A$. (We are considering here a pure thermal exchange where no mechanical friction or motion arises.) The decrease of entropy of the reference substance A is therefore $\Delta Q_i'/T_A$ and the increase of entropy of the reference substance B is $\Delta Q_i'/T_B$. Overall therefore the thermal energy transfer has given rise to a net *increase* of entropy of amount

$$\Delta Q_i' \left[\frac{1}{T_B} - \frac{1}{T_A}\right].$$

This result has led to the enunciation of a law in which is believed to be completely valid in general, viz. that the entropy of the universe must always increase. We can see the reasons for this from both mechanical and thermal points of view as follows. All mechanical experience indicates to us that any real process will be accompanied by frictional resistance. This causes directly an increase of entropy in the reference substance actually experiencing the friction. We can if we wish offset that particular increase by the extraction of energy by the heating process but to do so we inevitably cause a greater increase of entropy in some other piece of matter.

Now while this principle of the general increase of entropy is important to physics and philosophy, the engineer must not be led thereby into an aura of mysticism about the property entropy. For him it is, like volume, energy, temperature, etc., a relevant property of a substance which he may be processing – and which he can increase or decrease at will. Naturally to cause any change in either direction of this property for a particular reference substance, as well as of any other, he must operate within the laws of nature. But he must in his thinking treat it as subject to his control just as he does other properties, and use it with equal facility as a conceptual tool.

To emphasise this point we repeat the earlier definition of an isentropic change as one in which the entropy of *the reference substance* is kept constant, i.e. on the same status as the definitions of isothermal and of isopiestic changes.†

However in order to make it a serviceable conceptual tool we have now to consider how we can in fact set up measurements or make statements about the value or quantity of entropy in a substance. The defining equation (4.1.7) is only a beginning. We proceed now to develop a further

† Unfortunately there is sometimes adopted an alternative viewpoint of defining an isentropic change as one which is both mechanically and thermally ideally reversible so that the entropy of all the interacting elements of the process is constant. This view removes the concept of isentropic change out of the vocabulary of practical affairs and can seriously limit conversation.

conceptual framework, which will include clarification of how the thermodynamic temperature T, defined conceptually in Chapter 3, may in principle be measured.

4.2 Other thermodynamic properties and their relations

To our earlier recognition of p, V, T and U as thermodynamic properties, we have now added the recognition of the property S. It is interesting to note in passing the relative status of these. The first three arise directly in our phenomenological experience, even if temperature is much harder to rationalise than are p and V. Internal energy U is an idea extended from our macroscopic experience to a belief in energy inhering in atomic and molecular particles. Entropy S is deduced as a property which must exist, given the first and second laws of thermodynamics.

We have already seen that at least one other property of interest could be recognised in the combination $U+pV$, forming the enthalpy H. This combination property is exceedingly useful in considering any flow process, since we saw in Chapter 2 that, where bulk kinetic energy and potential energy changes are negligible,

$$\Delta Q_t - \Delta W_d = dH. \tag{4.2.1}$$

It is also found useful to define other combination properties. The following are of particular interest.

(*a*) *Free energy* (sometimes called Helmholtz free energy, or Helmholtz function, or free energy at constant volume).

$$F = U - TS. \tag{4.2.2}$$

(*b*) Gibbs Function (sometimes called Gibbs free energy, or free energy at constant pressure)

$$G = H - TS. \tag{4.2.3}$$

The differentials of the combination properties are:

$$dH = dU + p\,dV + V\,dp = T\,dS + V\,dp; \tag{4.2.4}$$

$$dF = dU - T\,dS - S\,dT = -p\,dV - S\,dT; \tag{4.2.5}$$

$$dG = dH - T\,dS - S\,dT = V\,dp - S\,dT. \tag{4.2.6}$$

From (4.2.5) we have

$$\left(\frac{\partial F}{\partial V}\right)_T = -p; \tag{4.2.7}$$

$$\left(\frac{\partial F}{\partial T}\right)_V = -S. \tag{4.2.8}$$

The suffix denotes a property which is kept constant during the change.

Now differentiate (4.2.7) with respect to T keeping V constant, and we have

$$\frac{\partial^2 F}{\partial T\,\partial V} = -\left(\frac{\partial p}{\partial T}\right)_V. \tag{4.2.9}$$

Similarly we differentiate (4.2.8) with respect to V keeping T constant and find

$$\frac{\partial^2 F}{\partial V\,\partial T} = -\left(\frac{\partial S}{\partial V}\right)_T. \tag{4.2.10}$$

Hence from (4.2.9) and (4.2.10) we have

$$\left(\frac{\partial p}{\partial T}\right)_V = \left(\frac{\partial S}{\partial V}\right)_T. \tag{4.2.11}$$

By a similar successive differentiation of (4.2.6) we obtain

$$\left(\frac{\partial V}{\partial T}\right)_p = -\left(\frac{\partial S}{\partial p}\right)_T. \tag{4.2.12}$$

Equations (4.2.11) and (4.2.12) are two of a set of four relations between thermodynamic properties known as *Maxwell's relations*. The others do not concern us at present. Similarly, although the functions F and G are of considerable use in other connections and will be discussed later, we have used them at the present stage merely to derive equations (4.2.11) and (4.2.12), which are essential for our immediate concern of establishing how the thermodynamic temperature may be measured.

The clue comes from equation (4.2.11). Recalling that

$$dS = \frac{dU + p\,dV}{T},$$

we have

$$\left(\frac{\partial S}{\partial V}\right)_T = \frac{1}{T}\left[p + \left(\frac{\partial U}{\partial V}\right)_T\right]. \tag{4.2.13}$$

Hence, using (4.2.11) we obtain

$$\frac{1}{p}\left(\frac{\partial V}{\partial T}\right)_V = \frac{1}{T}\left[1 + \frac{1}{p}\left(\frac{\partial U}{\partial V}\right)_T\right]. \tag{4.2.14}$$

But the quantity $1/p(\partial p/\partial T)_V$ is the coefficient of pressure variation with temperature at constant volume, commonly denoted by β, and this is a quantity which is relatively easy to measure. From (4.2.14) we see that

$$T = \frac{1}{\beta}\left[1 + \frac{1}{p}\left(\frac{\partial U}{\partial V}\right)_T\right]. \tag{4.2.15}$$

Hence if we can find a substance for which $(\partial U/\partial V)_T$ is zero, i.e. for which the internal energy at constant temperature is independent of the volume, we can then measure the thermodynamic temperature exactly, simply by measuring the value of β. Consider how this would operate. Suppose we are using the Celsius degree as our arbitrary unit and want to find the thermodynamic temperature, in such units, of a particular piece of matter at some definite state. Our thermometer will consist in principle of a closed volume containing the substance for which $(\partial U/\partial V)_T$ is zero, and we shall insert it into the reference substance at temperature T and note the pressure when in equilibrium. Now we make the temperature increase by a known small interval in terms of our Celsius degrees. Let us say we make it change by one-tenth of a Celsius degree. We note the increase of pressure in our thermometer. Suppose it has increased by a fraction 0.001 25 of the previous value. Then

$$\frac{\Delta p}{p} = 0.001\,25 \quad \text{and} \quad \beta = \frac{1}{p}\frac{\Delta p}{\Delta T} = \frac{0.001\,25}{0.1} = 0.0125.$$

Hence $T = 1/\beta = 80$ Celsius degree units. There is clearly an element of uncertainty of the order of 0.1 Celsius degrees in the example given, but if the pressure measurement can give smaller changes with accuracy, there is no difficulty in *principle* in increasing the accuracy of the measurement of the thermodynamic temperature.

Thus we see that despite the apparent abstract nature of the thermodynamic temperature, there is an effectively simple way of obtaining actual values for it, provided we use a particular class of substance, i.e. that for which $(\partial U/\partial V)_T = 0$. We now proceed to discuss the practicality of obtaining such substances.

4.3 The perfect gas and thermodynamic temperature

It has long been known that all gases, when the pressure is not too high, conform approximately to Boyle's law – i.e. that the product pV, the flow function, is constant at constant temperature. We define a *perfect gas* as one which obeys Boyle's law precisely. Thus the product pV depends neither on pressure nor volume, but is a function of temperature only. Let us consider the consequences which follow from this.

We have

$$pV = f(T), \tag{4.3.1}$$

$$\therefore \quad \beta = \frac{1}{p}\left(\frac{\partial p}{\partial T}\right)_V = \frac{1}{pV}f'(T) = \frac{f'(T)}{f(T)}. \tag{4.3.2}$$

4.3. THE PERFECT GAS

In addition to β, the coefficient of pressure variation with temperature, it is useful to define the coefficient of expansion, denoted by α, as

$$\alpha = \frac{1}{V}\left(\frac{\partial V}{\partial T}\right)_p. \tag{4.3.3}$$

Hence for a perfect gas we find also

$$\alpha = \frac{1}{Vp}f'(T) = \frac{f'(T)}{f(T)}. \tag{4.3.4}$$

Thus a perfect gas has the property that

$$\alpha = \beta, \tag{4.3.5}$$

and is a function of T only.

Now we recall, from equation (4.2.12), that, for any substance

$$\left(\frac{\partial V}{\partial T}\right)_p = -\left(\frac{\partial S}{\partial p}\right)_T, \tag{4.3.6}$$

$$\therefore \quad \alpha = -\frac{1}{V}\left(\frac{\partial S}{\partial p}\right)_T. \tag{4.3.7}$$

Substitution for $\qquad dS = \dfrac{dU + p\,dV}{T}$

gives $\qquad \left(\frac{\partial S}{\partial p}\right)_T = \frac{1}{T}\left[p\left(\frac{\partial V}{\partial p}\right)_T + \left(\frac{\partial U}{\partial p}\right)_T\right], \qquad (4.3.8)$

$$\therefore \quad \alpha = -\frac{1}{T}\left[\frac{p}{V}\left(\frac{\partial V}{\partial p}\right)_T + \frac{1}{V}\left(\frac{\partial U}{\partial p}\right)_T\right]. \tag{4.3.9}$$

Now from the general properties of partial differentials

$$\left(\frac{\partial V}{\partial p}\right)_T = -\frac{\left(\frac{\partial V}{\partial T}\right)_p}{\left(\frac{\partial p}{\partial T}\right)_V}. \tag{4.3.10}$$

Hence $\qquad \dfrac{p}{V}\left(\dfrac{\partial V}{\partial p}\right)_T = -\dfrac{\alpha}{\beta}. \qquad (4.3.11)$

Substitution in equation (4.3.9) gives

$$\alpha = \frac{1}{T}\left[\frac{\alpha}{\beta} - \frac{1}{V}\left(\frac{\partial U}{\partial p}\right)_T\right]. \tag{4.3.12}$$

Equations (4.3.6) to (4.3.12) are true for any substance. For a perfect gas, $\alpha = \beta$, so that (4.3.12) becomes

$$\alpha = \frac{1}{T}\left[1 - \frac{1}{V}\left(\frac{\partial U}{\partial p}\right)_T\right]. \tag{4.3.13}$$

And we had previously, equation (4.2.15),

$$\beta = \frac{1}{T}\left[1 + \frac{1}{p}\left(\frac{\partial U}{\partial V}\right)_T\right]. \tag{4.3.14}$$

We have already seen that for a perfect gas, α and β are not only equal, but must be a function of temperature only. Hence from (4.3.13) and (4.3.14)

$$\left.\begin{aligned} \frac{1}{V}\left(\frac{\partial U}{\partial p}\right)_T &= \phi(T) \quad \text{or} \quad \text{zero} \\ \frac{1}{p}\left(\frac{\partial U}{\partial V}\right)_T &= -\phi(T) \quad \text{or} \quad \text{zero} \end{aligned}\right\}, \tag{4.3.15}$$

where $\phi(T)$ is a function of temperature only.

Now suppose that $\phi(T)$ is not zero, then we should have

$$\left(\frac{\partial U}{\partial p}\right)_T = V\phi(T) = \frac{f(T)\,\phi(T)}{p}. \tag{4.3.16}$$

This can be integrated to give

$$\begin{aligned} U &= f(T)\,\phi(T)\log p + \text{constant with respect to pressure} \\ &= f(T)\,\phi(T)\log p + \psi(T) &(4.3.17) \\ &= f(T)\,\phi(T)\,[\log f(T) - \log V] + \psi(T). &(4.3.18) \end{aligned}$$

Equation (4.3.18) automatically meets the requirement of the second part of (4.3.15).

Now both (4.3.17) and (4.3.18) show that, if $\phi(T)$ is not zero, it would be possible for U to become negative, either below a certain pressure or above a certain volume, unless $\psi(T)$ were infinite for all values of temperature. Hence the only acceptable condition for (4.3.15), if we accept the view that U must always be an inherently positive quantity, is that both $(\partial U/\partial p)_T$ and $(\partial U/\partial V)_T$ are zero.

Hence the condition

$$pV = \text{constant at constant temperature},$$

the defining condition for a perfect gas, is sufficient to imply that the internal energy U is also a function of temperature only. Hence for a perfect gas

$$\alpha = \beta = \frac{1}{T}. \qquad (4.3.19)$$

It follows that any real gas, used within a range of pressure for which pV is a function of temperature only, may be used as a basis for a thermometer to measure thermodynamic temperature, as discussed in the previous section.

By this means it has been established that the thermodynamic temperature of the freezing point of pure water at standard atmospheric pressure is 273.15 when the unit interval is the Celsius unit – i.e. one-hundredth of the interval between the freezing and boiling points of pure water at standard atmosphere. Naturally it is $\frac{180}{100} \times 273.15 = 491.67$ when the unit interval is the Fahrenheit unit. Hence the thermodynamic temperature is found in Celsius units by adding 273.15 to the Celsius temperature, or in Fahrenheit units by adding 459.67 (i.e. $491.67 - 32$) to the Fahrenheit temperature. The former result is called the thermodynamic (or absolute) temperature, expressed in kelvins (symbol K) or alternatively, though less commonly now, in degrees Kelvin (symbol °K); while the latter is said to be in rankines (symbol R) or in degrees Rankine (symbol °R).† These notations and names are given because the reader will still see many textbooks which use them although under the S.I. Unit system the internationally agreed unit of thermodynamic temperature is now named 'kelvin' and denoted simply by the symbol 'K'.

Before concluding this section, we draw attention to another property of the perfect gas as defined. We have proved that for such a gas

$$\alpha = \frac{1}{V}\left(\frac{\partial V}{\partial T}\right)_p = \frac{1}{T}. \qquad (4.3.20)$$

Integration gives

$$\log T = \log V + \text{constant with respect to volume but may be a function of pressure.} \qquad (4.3.21)$$

Similarly, since

$$\beta = \frac{1}{p}\left(\frac{\partial p}{\partial T}\right)_V = \frac{1}{T}, \qquad (4.3.22)$$

$$\log T = \log p + \text{constant with respect to pressure but may be a function of volume.} \qquad (4.3.23)$$

† It is of some passing interest to note that Kelvin, who was born William Thomson, adopted for his title the name of a small river, the Kelvin, which winds by the campus of the University of Glasgow. Thus it is that the ancient name of a minor Scottish river becomes the name of a temperature unit.

Both (4.3.21) and (4.3.23) can only be satisfied if

$$\log p + \log V = \log T + \text{constant}.$$

Hence
$$pV = (\text{constant})\, T. \tag{4.3.24}$$

Thus we have proved that, given the defining condition that pV is a function of temperature only (equation (4.3.1)), the arguments, which are based on the laws of thermodynamics, lead to the conclusion that the only possible function is that pV is proportional to the thermodynamic temperature. The constant of proportionality is usually denoted by R, and we have

$$pV = RT.$$

In our nomenclature V is the volume of unit mass. Now it is known experimentally that for gases the volume of unit mass is inversely proportional to the molecular weight. Hence if V_m denotes the volume of one mole of gas, V_m is the same for all gases at given conditions of pressure and temperature.

But
$$pV_m = pMV = MRT. \tag{4.3.26}$$

Defining $R_m = MR$ as the gas constant per mole gives

$$pV_m = R_m T \tag{4.3.27}$$

as a universal equation applicable to any substance for the range over which it behaves as a perfect gas. The value of R_m is found to be 8.3143 kJ mole^{-1} K^{-1} or 1.986 Btu mole^{-1} R^{-1}.†

4.4 Thermal capacity

Consider a substance undergoing the process path shown in Figure 4.4.1.

Let ΔQ_t be the amount of energy transmitted to unit mass by heating for the element path from p, T to $p+dp$, $T+dT$ as shown. Then the ratio $(\Delta Q_t/dT)_a$ represents the quantity of thermal energy transmission required to raise the temperature from T to $T+dT$ under the conditions of the process path a.

If the process change is made to occur slowly with no mechanical disturbance, the ratio becomes $T(dS/dT)_a$ and is therefore a property of the substance *for the path element a*. This property is called the thermal

† N.B. An alternative and less complicated argument to identify the perfect gas temperature with the thermodynamic temperature is the usual one of performing a Carnot cycle on a perfect gas. This may certainly be used in elementary classes, but it has the logical disadvantage that the concept of thermal capacity is required, *and* an assumption of *constant* thermal capacity.

capacity per unit mass for the path a. For historical reasons it is most frequently called the *specific heat*.

Two particular process paths are of special interest, viz. constant volume and constant pressure. We define the specific thermal capacity at constant volume as

$$c_v = T\left(\frac{dS}{dT}\right)_V, \tag{4.4.1}$$

and at constant pressure as

$$c_p = T\left(\frac{dS}{dT}\right)_V. \tag{4.4.2}$$

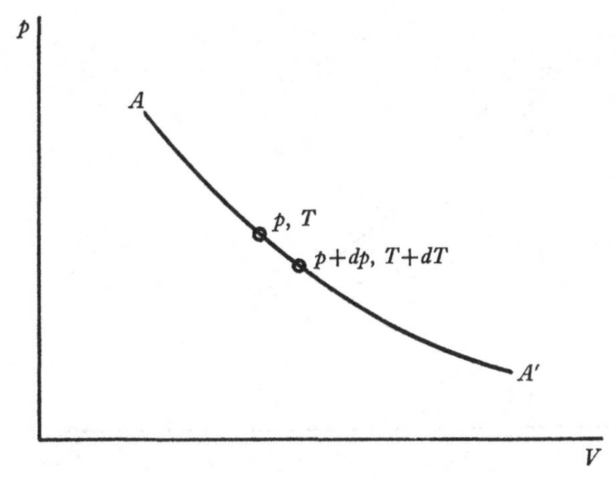

Fig. 4.4.1

Since $T\,dS = dU + p\,dV$, the following relations also apply:

$$c_a = \left(\frac{dU}{dT}\right)_a + p\left(\frac{dV}{dT}\right)_a; \tag{4.4.3}$$

$$c_v = \left(\frac{dU}{dT}\right)_V = \left(\frac{\partial U}{\partial T}\right)_V; \tag{4.4.4}$$

$$c_p = \left(\frac{\partial U}{\partial T}\right)_p + p\left(\frac{\partial V}{\partial T}\right)_p. \tag{4.4.5}$$

Since ΔQ_t can be measured in carefully controlled experiments under either constant volume or constant pressure conditions, values of c_v and c_p are easily obtainable in principle, and are characteristic properties of a substance. In general the values depend on the state of the substance.

There is one particular process path for which the ratio becomes an unacceptable concept – this is a process conducted at constant temperature. For such a process a more useful idea is to think of the quantity of thermal energy transmission required to increase the volume by one unit, i.e. the ratio $\Delta Q_t/dV$. This ratio is called the latent thermal capacity (or latent heat) of isothermal volume change and may be denoted by l_T. Hence we have

$$l_T = \frac{\Delta Q_t}{dV} = T\left(\frac{\partial S}{\partial V}\right)_T. \tag{4.4.6}$$

Now let us consider the small change from a to a' shown in Figure 4.4.2.

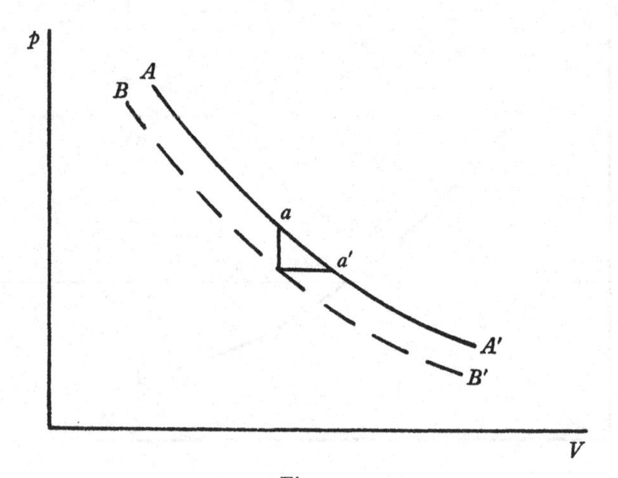

Fig. 4.4.2

Let AA' and BB' be two isothermal paths at respective temperatures $T+dT$ and T. The entropy change from a to a' may be made either

(i) along the path aa' for which

$$dS = \frac{l_T}{T}dV; \tag{4.4.7}$$

(ii) along the path aba' for which

$$dS = -c_v\frac{dT}{T}+c_p\frac{dT}{T}. \tag{4.4.8}$$

Since entropy is a property the change must be independent of path. Hence we obtain the important equation

$$c_p - c_v = l_T\left(\frac{\partial V}{\partial T}\right)_p. \tag{4.4.9}$$

But inserting (4.2.11) into (4.4.6) gives

$$l_T = T\left(\frac{\partial p}{\partial T}\right)_V;$$

$$\therefore \quad c_p - c_v = T\left(\frac{\partial p}{\partial T}\right)_V \left(\frac{\partial V}{\partial T}\right)_p \tag{4.4.10}$$

$$= T\alpha\beta p V. \tag{4.4.11}$$

Equation (4.4.11) is an important relation giving the difference between the specific thermal capacity at constant pressure and constant volume, for any substance, in terms of the temperature, the pressure, volume, and the expansion and pressure coefficients.

Substituting the values of α and β for a perfect gas gives

$$c_p - c_v = \frac{T.pV}{T^2} = R. \tag{4.4.12}$$

Hence the difference between the thermal capacities c_p and c_v for a perfect gas is constant, the value being equal to that of the gas constant.

Finally, since $(\partial S/\partial V)_T = (\partial p/\partial T)_V$, we may differentiate this relation with respect to T, keeping V constant, and obtain

$$\frac{\partial^2 S}{\partial T\,\partial V} = \left(\frac{\partial^2 p}{\partial T^2}\right)_V, \tag{4.4.13}$$

i.e.

$$\frac{\partial}{\partial V}\left(\frac{\partial S}{\partial T}\right) = \left(\frac{\partial^2 p}{\partial T^2}\right)_V; \tag{4.4.14}$$

$$\therefore \quad \frac{\partial}{\partial V}\left(\frac{c_v}{T}\right) = \left(\frac{\partial^2 p}{\partial T^2}\right)_V; \tag{4.4.15}$$

$$\therefore \quad \left(\frac{\partial c_v}{\partial V}\right)_T = T\left(\frac{\partial^2 p}{\partial T^2}\right)_V. \tag{4.4.16}$$

But for a perfect gas $(\partial p/\partial T)_V = R/V$, and hence with V constant the second derivative is zero. Thus $(\partial c_v/\partial V)_T$ is zero for a perfect gas – i.e. c_v is either a constant or a function of temperature only.

We had of course already established that U for a perfect gas is a function of temperature only, and hence the above conclusion regarding c_v tells us nothing more. Hence none of these arguments can lead to establishing a definite function of temperature for U. This will be understood when we recall that in (4.3.17) and (4.3.18) of the previous section, since $\phi(T)$ had to be zero, U becomes $\psi(T)$ which is an integration constant with respect to volume and pressure – i.e. it is a boundary condition which cannot itself be implied in the relation between p, V and T which defined the perfect gas.

However, it is also a well-established experimental observation that for all gases in the range of near perfect behaviour, c_v is in fact nearly constant. Hence it is usual to add as an additional specification defining the perfect gas the condition that c_v is a constant. (It is then really a pluperfect gas!) This enables some very simple relations to be obtained, which are sufficiently accurate for many purposes, provided a technique of using them is developed. Such techniques are much used for general engineering predictions and practice, and will be found in appropriate texts. However in general for real substances the thermal capacities do in fact depend on the temperature and pressure, as do in fact all the other characteristic properties.

4.5 Latent heat, temperature, and pressure in phase change

We know from everyday experience with water that as we transmit energy to it by heating, its temperature rises at atmospheric pressure until it reaches 100° C. No subsequent rise is noted, until all the liquid has turned into the gas which we call steam. Yet we have to continue supplying energy. Similar observations can be made on many substances.

This is therefore a particular example of a change occurring at constant temperature as discussed in Section 4.4 of this chapter. The concept of latent thermal capacity (or latent heat) is therefore useful. In the particular case of phase change, however, it is not customary to use the l_T of Section 4.4, but to use the concept of the amount of energy required to change unit mass of substance from one phase to the other.

It is also observed that if the substance is maintained at a higher pressure, it has to be raised to a higher temperature before the phase change occurs, and conversely at a lower pressure the phase change sets in at a lower temperature. The liquid is therefore said to have a ***vapour pressure*** which depends upon the temperature. Hence the evaporation occurs not only at constant temperature but also at constant pressure. We can show the sequence from liquid to vapour most clearly on a temperature/volume diagram as in Figure 4.5.1.

The lines marked p_1 and p_2 show constant pressure processes at two different pressures. In the liquid phase, heating causes rise of temperature which may be accompanied by small expansion until the 'boiling point', called saturation temperature, T_b is reached. This remains constant while the large expansion to the gas phase occurs. Once the substance is all gas, further heating again raises the temperature.

Denoting the entropy of unit mass of the gas phase at T_b, p by S_g and that of the liquid phase at T_b, p by S_f, it is evident that the energy transmission required to effect the change is given by:

$$\text{Latent heat} = \int \Delta Q_t$$

$$= \int T_b\, dS = T_b(S_g - S_f). \qquad (4.5.1)$$

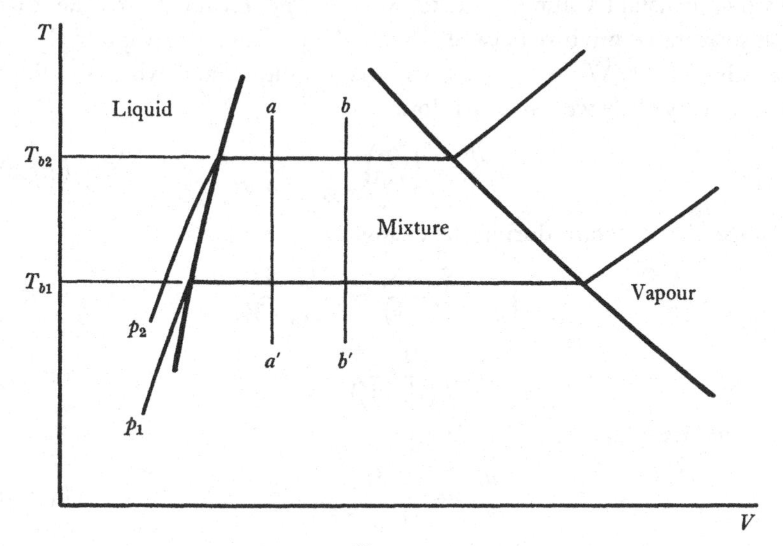

Fig. 4.5.1

Now consider the change in enthalpy:

$$dH = d(U+pV) = dU + p\,dV + V\,dp$$

$$= T\,dS + V\,dp. \qquad (4.5.2)$$

For the phase change, pressure is constant. Hence the latent heat of phase change is expressible as the difference in enthalpy per unit mass of each phase. It is now therefore conventional to denote the latent heat of the liquid/vapour phase change by the symbol H_{fg}, and we have

$$H_{fg} = H_g - H_f = T_b(S_g - S_f). \qquad (4.5.3)$$

Similarly for the phase change which occurs from solid to liquid in melting

$$H_{sf} = H_f - H_s = T_m(S_f - S_s). \tag{4.5.4}$$

An important equation expresses a functional relationship between the latent heat of phase change, the pressure, and the temperature. This is known as *Clapeyron's equation* and can be derived directly from the Maxwell relation of equation (4.2.11), viz.

$$\left(\frac{\partial p}{\partial T}\right)_V = \left(\frac{\partial S}{\partial V}\right)_T. \tag{4.5.5}$$

The vertical lines aa' and bb' in Figure 4.5.1 show that irrespective of which constant volume we take, we are at p_2, T_{b2} and p_1, T_{b1}, i.e. that in any saturated mixture between the total liquid and total vapour phases the value of $\Delta p/\Delta T_b$ at constant volume is unique, and when the lines become very close we have therefore

$$\frac{dp}{dT_b} = \left(\frac{\partial p}{\partial T}\right)_{V,\,\mathrm{sat}}. \tag{4.5.6}$$

Since T_b is constant during the change,

$$\left(\frac{\partial S}{\partial V}\right)_T = \frac{S_g - S_f}{V_g - V_f} = \frac{H_g - H_f}{T_b(V_g - V_f)}$$

$$= \frac{H_{fg}}{T_b(V_g - V_f)}. \tag{4.5.7}$$

Using (4.5.6) and (4.5.7) in (4.5.5) gives

$$\frac{dp}{dT_b} = \frac{H_{fg}}{T_b(V_g - V_f)}, \tag{4.5.8}$$

which is Clapeyron's equation.

Since V_g is always $> V_f$, dp/dT_b is always positive, i.e. boiling point or saturation temperature always increases with pressure.

Similarly for melting we have

$$\frac{dp}{dT_m} = \frac{H_{sf}}{T_m(V_f - V_s)}. \tag{4.5.9}$$

For some substances, water for example, $V_s > V_f$, i.e. it contracts on melting.

Nevertheless energy is required to melt the solid i.e. H_{sf} is positive. Hence for water dp/dT_m is negative, i.e. the freezing point diminishes as pressure is increased. Hence at a fixed temperature increase of pressure will melt ice.

4.6 Critical temperature and pressure

To conclude this discussion of characteristic thermodynamic properties of substances we may note briefly that we do not always find the behaviour of phase change discussed in the previous section. Figure 4.6.1 indicates how this occurs. It can best be understood by comparison with Figure 4.5.1. In Figure 4.6.1 the region boundary between the liquid/mixture and mixture/vapour which appears in Figure 4.5.1 is continued and we see that it has a turning point at a particular value of temperature, denoted by T_c. Hence at any temperature above T_c, no condition of pressure and volume can be in the liquid/vapour mixture region.

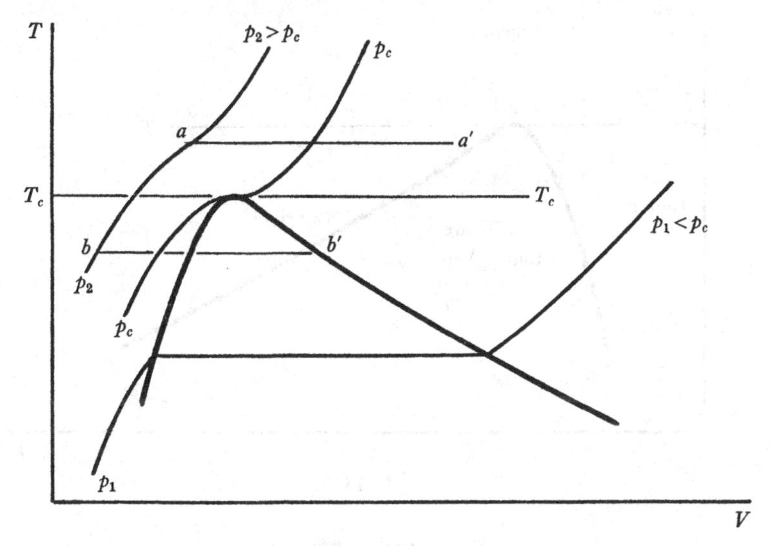

Fig. 4.6.1

One particular constant pressure line, denoted by p_c, is just tangent to the boundary curve at T_c.

We have already seen the behaviour with a pressure such as p_1 which is $< p_c$. Consider now a pressure which is greater than p_c, such as p_2. If we raise the temperature, it passes through T_c without any discontinuity of volume along the path *ba*. There is no marked change between the 'liquid' phase and the high pressure 'gas' phase above T_c. Nor if we reduce pressure from the point *a* along the line *aa'* is there any discontinuity. If however we stopped raising the temperature at the point *b*, and, maintaining the temperature, which is less than T_c, constant, reduced the pressure progressively we would move along the line bb', going through a mixture discontinuity before reaching the vapour phase. Thus there is never any discontinuity of phase as long as the temperature is above T_c – or, as is frequently said – it is impossible to liquefy a gas, no matter how high the pressure, so long as the temperature is above T_c. T_c is therefore called the critical temperature, and the tangent pressure p_c is similarly called the critical pressure.

The regions in the T, V diagram can therefore be considered to have significance as shown in Figure 4.6.2.

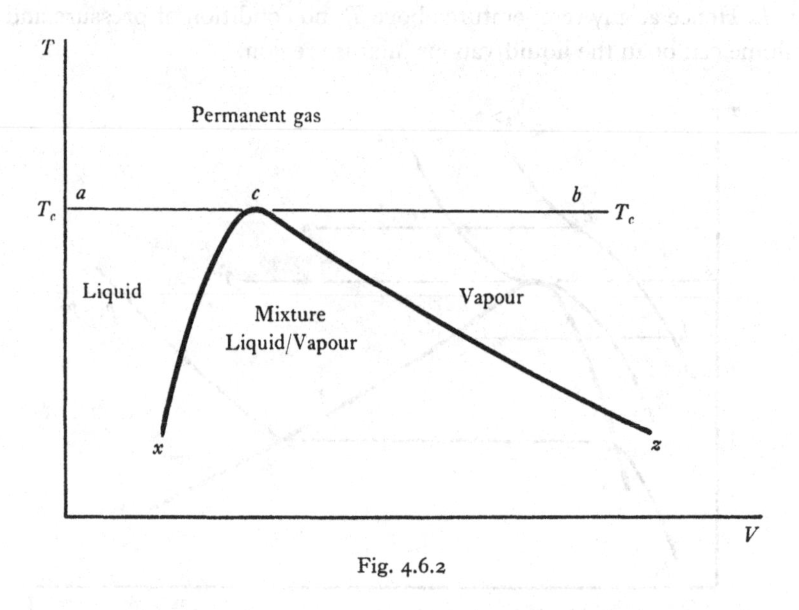

Fig. 4.6.2

The region within *acx* is liquid, that within *bcz* is vapour, and within *xcz* is liquid/vapour mixture. The region above *acb* is permanent gas. There is no discontinuity of properties in crossing the line *ac* from liquid to gas or vice versa, nor in crossing the line *cb* from vapour to gas or vice versa. We can think easily of the vapour and gas as being similar, and have no difficulty in imagining that crossing *cb* is continuous. We must be equally clear that at pressures above the critical crossing the line *ac* is also continuous.

The critical pressure and temperature are very different for different substances. Water, for example, has $T_c = 647\,\text{K}$ (374° C) with $p_c \doteqdot 21.8\,\text{MN/m}^2$ (218 atmospheres). Oxygen has $T_c = 154\,\text{K}$ (−119° C) with $p_c \doteqdot 4.97\,\text{MN/m}^2$ (49.7 atmospheres). Hence we experience oxygen in the air as a permanent gas and have to go to considerable trouble and expense to liquefy it, while water is present in the air as vapour, frequently condensing as rain, and lying in the 'pools' of seas and lakes, from which it evaporates, i.e. we experience water as a mixture of liquid and vapour under atmospheric conditions because our ambient temperatures are all far below T_c for water.

Against this background of critical temperature and pressure we can now appreciate more fully the conditions under which we may expect a gas to have an equation of state approximating to that of a perfect gas. It is found that the main requirement is for the pressure to be well below p_c, i.e. where previously we said that many gases behaved as near perfect gases at low pressure, what is really meant is that the pressure must be low *compared with* p_c. Provided this is the case, many substances approximate quite well to $pV = RT$ not only as gases but also as vapours, i.e. even when the temperature is well below T_c.

This is a convenient fact of which we can make use in practice.

4.7 Practical enumeration of properties

For practical purposes we require to have arithmetical values to give to all the various thermodynamic properties of a substance. It is apparent from first principles that this can be done easily in absolute terms for pressure and volume, since these are enumerated in terms of units of force and of distance, and in each such case we know definitely when there is zero force and zero distance. Further it has been shown by thermodynamic argument that we can enumerate temperature in an absolute fashion since again we know a zero. Hence for the properties p, V, and T we have no difficulties in principle in giving numerical values. But the situation is very different when we come to the properties U and S.

From the previous work it is apparent that in an appropriately planned experiment we can measure *changes* of U and of S. Thus for example from the relation

$$\Delta Q_t - \Delta W_t = dU \tag{4.7.1}$$

we can by an adiabatic experiment, i.e. with $\Delta Q_t = 0$ throughout, obtain the change $U_2 - U_1$ by measuring mechanically the energy transmitted by the work process. For we have

$$U_2 - U_1 = -\int_1^2 \Delta W_t. \tag{4.7.2}$$

Alternatively, we might do an experiment where both

$$\int_1^2 \Delta Q_t \quad \text{and} \quad \int_1^2 \Delta W_t$$

were measured, or do an experiment heating at constant volume with no friction where

$$\int_1^2 \Delta Q_t = U_2 - U_1.$$

Again from the relation

$$dS = \frac{dU + p\,dV}{T} \tag{4.7.3}$$

we can by a suitable experiment measure $\int dU$ as above and also $\int p\,dV$, and the temperature throughout and so obtain a numerical value of the entropy *change* as

$$S_2 - S_1 = \int_1^2 \frac{dU}{T} + \int_1^2 \frac{p\,dV}{T}. \tag{4.7.4}$$

But we have no *a priori* knowledge of what the *absolute* value of U or of S is in any piece of substance. Practical objectives of constructing tabulated values and graphs of these properties for engineering purposes have therefore to be achieved by taking some particular state of the substance and arbitrarily giving it the values $U = 0$ and $S = 0$. Then all other states can be referred to this by experiments using relations of the type (4.6.2) and (4.6.4), or based on these.

Once values of U and S have been determined in this way, we can also give numerical values to the composite property H from the relation

$$H = U + pV.\dagger \tag{4.7.5}$$

† *Warning.* It might be thought that similarly we could give numerical values to the other composite properties F and G mentioned in Section 4.2 from the relations

$$\left.\begin{aligned} F &= U - TS, \\ G &= H - TS. \end{aligned}\right\} \tag{4.7.6}$$

But this is not so because each of F and G has a *double* arbitrariness, in that both arbitrary property values U and S appear, whereas in H only the single arbitrary property value U occurs. As a result we have for H

$$dH = dU + p\,dV + V\,dp,$$

$$\therefore \quad dH = T\,dS + V\,dp. \tag{4.7.7}$$

Since p, V and T are all known absolutely, and since we *can* measure *changes* of S, it follows from (4.7.7) that we can usefully discuss changes of H. But when we come to F and G we have

$$\left.\begin{aligned} dF &= dU - T\,dS - S\,dT, \\ \therefore \quad dF &= -p\,dV - S\,dT. \\ dG &= dH - T\,dS - S\,dT, \end{aligned}\right\} \tag{4.7.8}$$

$$\therefore \quad dG = V\,dp - S\,dT. \tag{4.7.9}$$

4.7. PRACTICAL ENUMERATION OF PROPERTIES

Since we do not know the actual absolute value of S we cannot use equations (4.7.8) and (4.7.9) to discuss changes of F and G, except in cases where temperature is constant. Hence although we can compile tables of p, V, T, U, S, and H for a substance based on arbitrary specification of U and S to be zero at a particular state, and *differences* of U, S and H in such tables are all meaningful, and although we could formally operate arithmetically to get values of F and G, such values would be meaningless except for their *differences at identical temperature.*

In more advanced thermodynamics methods are developed to overcome this problem, but in general the student should take great care not to use numerical values of the properties F and G until he is able to satisfy himself that such techniques have been included and that the values are therefore meaningful.

It must of course be understood that these cautionary remarks on arithmetical values do not in any way detract from the utility of using the concepts F and G for theoretical development and manipulation.

5

REVIEW AND SUMMARY OF CHAPTERS 1–4. THE MISSING ITEM

You've got the snaffle and the bit all right.
But where's the bloody horse?

Roy Campbell

We have now reached a stage in our discussion where it is profitable to stop temporarily, to survey the sequence of thought from the beginning, and to take stock of what has been achieved.

We began with the general laws of mechanics leading to the concept of energy and of its conservation. From this there emerged the concept of internal energy, and we found that the existence of thermal energy transfer in addition to energy transfer by work was predicted. This thermal phenomenon is characterised in our experience by the parameter temperature and the laws had to include the experience of the unidirectional flow of thermal energy in relation to temperature. This led to a basic law limiting the extent of convertibility of thermal energy transmission to work transmission. From this there emerged an absolute significance for temperature, and an absolute – if rather abstract – definition of T, thermodynamic temperature. This opened the way to showing that the property entropy S must exist in addition to the directly observable properties p, V, θ and basically assumed property U with which we started.

Chapter 4 continued to set up a scheme of relations between these properties and other definable characteristics. Thus beginning in Chapter 1 with *general* laws of *process*, we concluded in Chapter 4 with specific characteristics of *substances*, and inter-relations between these characteristics.

Moreover, we found that a particular class of substance, the perfect gas, can in principle be used as a thermometer to measure the thermodynamic temperature. Since real gases for certain conditions behave as near perfect, the thermodynamic temperature was taken from the realm of abstraction into that of concrete practicality and utility.

We are therefore now furnished with a scheme of properties, relations, and characteristics of actual substances, all of which are defined and consonant with our basic assumptions and the general laws of mechanical and thermodynamic process. Many other such relations can be established and detailed texts in physical thermodynamics properly derive these and discuss their significance and application much further. Similarly the relations established in Chapters 1 and 2 for dynamic work and net or total work are developed in specialist texts in engineering thermodynamics to assist in the understanding, the analysis, and the design, of thermodynamic machines such as engines, turbines, compressors, pumps, etc. Special techniques for such analysis are derived in such texts to make use of the properties thermal capacity, enthalpy, etc., etc. In this book we shall not attempt to deal with these techniques, since our objective is a general introduction giving an overall perspective.

From the above survey of what we have obtained so far, it is apparent that one very important item is still missing. For while we have established clearly that energy supply by the process of heating is essential to the production of work in a uniform pressure and stationary environment devoid of potential, we have no indication yet of how such energy supply may be obtained. Admittedly we have recognised a necessary condition that there must be a source at higher temperature. But of how we can procure such higher temperature and have energy supply from it our *logical* structure has so far had nothing to say. Of course our practical knowledge tells us that we obtain high temperatures and energy supply by the process which we know as burning, or combustion. Our thermodynamics has told us how this can be used, but not how it can arise. The next matter to which we turn attention is therefore the study of what thermodynamics can tell us about how energy supply at temperatures above ambient does in fact occur. We proceed to this matter in the next chapter, where we discover that it is one aspect of the thermodynamics of chemical change.

But as a preliminary we remind the reader that while in Section 2.6 of Chapter 2, attention was drawn to the fact that the specific volume occupied by unit mass of a substance depends on three things – (*a*) its chemical nature, (*b*) its state of stress, and (*c*) its temperature; our study up to now has been concerned with (*b*) and (*c*) only, (*b*) being characterised by the pressure, so that we have been dealing entirely with the concept of a single pure substance.

But the fact of different kinds of substance is as much a basic experience for us as are the facts of force (pressure), and of temperature. The 'happenings' which take place between two substances of different kinds are as obvious to us as the transmission processes of work and of heating. Thus we are familiar with such processes as corrosion and combustion in everyday life, and recognise that in such cases there are always present substances which are different in kind. We learn that substances of very different characteristics may interact to produce others which differ from any of the originals, and that a substance may decompose. We are also aware that transmission processes of work and of heating may accompany this change of species of substance. Hence we shall be interested to discover how the laws of thermodynamics as we have developed them will assist in the understanding, description, and prediction of this process which we call chemical change.

However, in our approach to chemical phenomena we shall try to repeat the same kind of argument as characterised the beginning of our studies. We shall pretend ignorance of chemical change, just as in Chapter 1 we pretended ignorance of thermal energy transmission. We demonstrated there that the existence of a mode of energy transmission other than that of work arose naturally from the argument and then identified this with thermal phenomena. We shall hope, therefore, that we can now develop, from the conceptual structure as we now have it, a prediction of some kind of behaviour which can be identified with chemical change.

6

PHASE CHANGE AND CHEMICAL PROCESSES

You can't swim in the same river twice

Anon., probably based on Heraclitus 'Everything Changes'

6.1 Basic conditions for extensivity

The general notion of extensivity can be defined most precisely in the following way. A characteristic will be said to be extensive when its value for a whole is equal to the sum of the values for *any* parts into which the whole may be conceptually divided. From this definition we see immediately that pressure and temperature are not extensive. We also see that specific volume, i.e. volume per unit mass, is not extensive – for the volume per unit mass of a whole mass is not obtained by summing the volume per unit mass of any set of parts making up the whole mass. The same is true of some other characteristics of unit mass such as specific internal energy and specific enthalpy. Such non-extensive characteristics are called intensive. In all the preceding chapters we have assumed that every subject under discussion could be dealt with simply in terms of a unit mass, and that an overall result for any given mass could be obtained in proportion to the mass. We now require to consider whether there are any circumstances under which this assumption can be invalid.

Our basic pattern of thought requires that we believe mass to be an extensive characteristic, i.e. that a whole mass is equal to the sum of the masses of its parts. Undoubtedly we also consider that volume behaves similarly – the volume of a whole is the sum of any set of part volumes making up the whole. We also believe the same to be true of energy. Thus if we denote extensive parameters by stressed symbols and intensive properties of unit mass in the usual unstressed symbols, we shall have for a mass M:

$$U' = \sum_i M_i U_i, \quad V' = \sum_i M_i V_i, \tag{6.1.1}$$

where suffix i refers to any part M_i of the total mass M and

$$\sum_i M_i = M. \tag{6.1.2}$$

But what about the other characteristics which we have developed and used in our discussion, namely enthalpy and entropy? Let us consider first of all the case of enthalpy. We shall have, if we *assume* extensivity,

$$H' = \sum_i M_i(U_i + p_i V_i) = \sum_i M_i U_i + \sum_i M_i p_i V_i;$$

$$\therefore \quad H' = U' + \sum_i M_i p_i V_i. \tag{6.1.3}$$

Unless p is uniform throughout the whole mass M we cannot give a general summation to equation (6.1.3). Provided p is uniform throughout M, however, we shall have

$$H' = U' + p\sum_i M_i V_i = U' + pV'. \tag{6.1.4}$$

Thus when p is uniform throughout the whole mass, and only when this is so, the enthalpy of the whole will have the same value when determined from the energy, pressure, and volume of the whole as from the sum of the individual enthalpies of the portions.

When we come to consider the entropy we are faced with the problem that we do not have any functional relation, in general, between S and U and V as we had for H. But we do have the relation for entropy change, viz. for unit mass,

$$T\,dS = dU + p\,dV. \tag{6.1.5}$$

If we assume that this applies extensively for each part mass, we have

$$T_i M_i dS_i = M_i dU_i + p_i M_i dV_i; \tag{6.1.6}$$

$$\therefore \quad \sum_i T_i M_i dS_i = \sum_i M_i dU_i + \sum_i p_i M_i dV_i. \tag{6.1.7}$$

Now from equations (6.1.1), if we regard each constituent part M_i as constant, we shall have

$$dU' = \sum_i M_i dU_i, \quad dV' = \sum_i M_i dV_i. \tag{6.1.8}$$

We have already seen that to have the enthalpy calculation for the sum of the parts agree with that of the whole we must have all p_i equal, i.e. p uniform throughout the whole mass. Using this, and equations (6.1.8) in (6.1.7) we find

$$\sum_i T_i M_i dS_i = dU' + p\,dV'. \tag{6.1.9}$$

We now see that if $\sum_i M_i dS_i$ is to be equal to dS' for the whole mass M we must have T_i equal in all part masses, i.e. T must be uniform throughout M.

Thus to provide that enthalpy H' and entropy S' shall each be extensive parameters consistent with the basic extensive concepts energy U' and volume V', we must have both intensive properties p and T uniform throughout the mass M. Then for the whole mass the relations

$$\left.\begin{aligned} dU' &= T\,dS' - p\,dV', \\ dH' &= T\,dS' + V'\,dp, \end{aligned}\right\} \qquad (6.1.10)$$

will apply.

These are simply proportionate extensions of the relations for unit mass

$$\left.\begin{aligned} dU &= T\,dS - p\,dV, \\ dH &= T\,dS + V\,dp. \end{aligned}\right\} \qquad (6.1.11)$$

Thus the conditions of extensivity of enthalpy and entropy established so far, given the assumption that internal energy and volume are inherently extensive, are that the intensive properties p and T shall be uniform. As was seen in Chapter 2 this constitutes equilibrium in a single phase single substance and so implies also that all specific properties are uniform.

6.2 A further condition for extensivity. The Gibbs function

In deriving equations (6.1.8) and (6.1.9) and (6.1.10) we assumed that the mass M_i of each part of the whole M was constant. But of course we may consider many different sets of subdivisions of M, and any relevant laws must be capable of applying to all possible sets. The only constraint† on the possible sets of subdivisions is that given in equation (6.1.2), which establishes that

$$\sum_i dM_i = 0. \qquad (6.2.1)$$

We must therefore now investigate the possibility of extensivity of the parameters H' and S' for any possible redistribution of mass between subdivisions subject only to the constraint (6.2.1). The extensivity of volume V' and of energy U' is taken as basically essential in our thought pattern, and as axiomatically true for all possible subdivisions. Hence we assume always

$$\left.\begin{aligned} U' &= \sum_i M_i U_i \\ V' &= \sum_i M_i V_i \end{aligned}\right\}. \qquad (6.2.2)$$

† At a later stage an *additional* constraint is added to equation (6.2.1) for the special case of chemical *reaction*. But the validity of equation (6.2.1) is still preserved, as is the whole discussion in this section.

Now, when the mass M_i may vary, we have

$$\begin{aligned} dU'_i &= d(M_i U_i) \\ &= M_i dU_i + U_i dM_i. \end{aligned} \tag{6.2.3}$$

But, from (6.1.11) for unit mass,

$$dU_i = T dS_i - p dV_i;$$

$$\therefore \quad dU'_i = TM_i dS_i - pM_i dV_i + U_i dM_i. \tag{6.2.4}$$

Also

$$\begin{aligned} dS'_i &= d(M_i S_i) \\ &= M_i dS_i + S_i dM_i, \end{aligned} \tag{6.2.5}$$

and

$$dV'_i = M_i dV_i + V_i dM_i. \tag{6.2.6}$$

Substituting from (6.2.5) and (6.2.6) into (6.2.4) for $M_i dS_i$ and $M_i dV_i$ we obtain

$$dU'_i = T dS'_i - p dV'_i + (U_i + pV_i - TS_i) dM_i; \tag{6.2.7}$$

$$\therefore \quad \sum_i dU'_i = T \sum_i dS'_i - p \sum_i dV'_i + \sum_i (U_i + pV_i - TS_i) dM_i. \tag{6.2.8}$$

But since U and V are basically extensive we must have

$$\sum_i dU'_i = dU',$$

$$\sum_i dV'_i = dV'.$$

Hence equation (6.2.8) gives

$$dU' = T \sum_i dS'_i - p dV' + \sum_i (U_i + pV_i - TS_i) dM_i. \tag{6.2.9}$$

Comparing this with the corresponding relation for dU' in (6.1.10) we see that

$$T \sum_i dS'_i = T dS' - \sum_i (U_i + pV_i - TS_i) dM_i. \tag{6.2.10}$$

Hence the entropy change dS' of the whole mass M is not equal to the sum of the entropy changes of its parts, even with uniform temperature and pressure, unless we also have

$$\sum_i (U_i + pV_i - TS_i) dM_i = 0. \tag{6.2.11}$$

Now also

$$\sum_i dH'_i = T \sum_i dS'_i + \sum_i V'_i dp = T \sum_i dS'_i + V' dp.$$

Hence *both* entropy change and enthalpy change will be fully extensive, with uniform pressure and temperature provided equation (6.2.11) is satisfied.

One condition under which this is satisfied is of course that assumed in Section 6.1, i.e. when the set of M_i is fixed with no re-distribution of mass between the parts. But another condition is possible. Because of the constraint (6.2.1), equation (6.2.10) will also be satisfied if $U_i + pV_i - TS_i$ is a constant for all i. Thus the condition for entropy, as well as energy, volume, and enthalpy, to be a purely extensive characteristic irrespective of any re-distribution of mass between all possible sub-divisions is that the intensive function $U + pV - TS$ per unit mass, in addition to temperature and pressure, must be uniform throughout the whole mass M.

This function $U + pV - TS$ of the properties of unit mass is clearly of importance and we shall realise this fully in the later discussion. Meanwhile we shall, for algebraic convenience, adopt for it a single symbol G and indicate later why this particular symbol is in almost universal use.

We have been led to two alternative conclusions. Alternative (*a*) implies that if G is not uniform for unit mass in all possible portions of the total mass M, the entropy change for the total can be the mere sum of the corresponding parameters for all possible portions within M, only if such portions all have boundaries impermeable to mass, so that no redistribution of mass between the sub-divisions can take place. Alternative (*b*) implies that if the total mass can be re-distributed between the sub-divisions, G must be uniform for unit mass throughout all possible portions of M. We should also note the following further implications.

(*c*) It is also implied by these arguments that if the various submasses M_i cannot be varied, i.e. if they have boundaries impermeable to mass, the entropy change for the total mass is identical to the sum of the entropy changes for all the parts, even if G_i is not equal in all parts.

(*d*) The final and most important point to discuss here is what is implied when neither alternative (*a*) nor (*b*) holds – i.e. what happens when re-distributions of mass between the portions can occur *and* G_i is not equal in all portions. From conclusion (*c*) it follows that when neither (*a*) nor (*b*) holds, the difference $\sum_i dS'_i - dS'$ is the difference in entropy change which occurs as the result of re-distribution of mass when G is not uniform. And from equation (6.2.10) we see that this entropy change difference is given by

$$T\sum_i dS'_i - T\,dS' = -\sum_i G_i\,dM_i. \tag{6.2.12}$$

Now we know that for any change to occur spontaneously, entropy must increase. Hence

$$T\sum_i dS'_i - T\,dS' > 0. \tag{6.2.13}$$

It follows that any spontaneous re-distribution of mass between subordinate portions of M must be in a direction such that

$$\sum_i G_i\,dM_i < 0. \tag{6.2.14}$$

If we wanted to make some re-distribution occur which would be of a nature that $\sum_i G_i\,dM_i > 0$ we should require to apply to M some operation. This kind of re-distribution would not occur spontaneously.

The kind of spontaneous re-distribution can be envisaged clearly in the simple case where we consider M to consist of two portions M_a and M_b, and the re-distribution to consist of a mass m transferring from M_a to M_b, so that $dM_a = -m$ and $dM_b = m$. For this simple case, relation (6.2.14) gives

$$m(G_b - G_a) < 0, \tag{6.2.15}$$

or

$$G_a - G_b > 0, \tag{6.2.16}$$

i.e. the mass m transfers *from* the portion with the higher value of G *to* the portion with the lower value of G.

Now since one cannot imagine that a re-distribution of this kind can continue unceasing, we must suppose that as a result of the transfer of mass from M_a to M_b the value of G_a per unit mass in M_a is reduced and of G_b per unit mass in M_b is decreased until eventually they become equal, when the process of re-distribution will stop.

Similarly for the case of many subordinate portions we must conclude that mass will in general be re-distributed *from* portions with higher values of G *to* those with lower values, until the values throughout become uniform.

What has been described is essentially a process by which an equilibrium is reached. By providing uniform pressure and temperature conditions we implied mechanical and thermal equilibrium. But we have seen now that if the function $G = U + pV - TS$ is not also uniform, there will be some other kind of disequilibrium and that a process of re-distribution of mass to reach equilibrium by establishing uniformity of G as well as of p and T will occur, with increasing entropy.

We turn now to the discussion of the energy changes which may accompany such re-distribution.

6.3 Energy transmission effects in re-distribution of mass

When a redistribution of mass occurs within a mass M containing $M_1, M_2, \ldots, M_i$, etc., the energy change is given by using equation (6.2.9),

$$dU' = T\sum_i dS'_i - p\,dV' + \sum_i G_i\,dM_i. \tag{6.3.1}$$

But from the conservation of energy assumption we also write

$$dU' = \Delta\psi', \tag{6.3.2}$$

where $\Delta\psi'$ represents the total transmission of energy to the mass M by any possible transmission process or processes. Assuming that only two modes of transmission, by heating and by work exist, we have with the usual conventions

$$\Delta\psi' = \Delta Q'_t - \Delta W'_t. \tag{6.3.3}$$

For any of the individual masses we shall have,

$$\Delta\psi'_i = \Delta Q'_{t,i} - \Delta W'_{t,i} \tag{6.3.4}$$

it being recognised that some of this transmission will be from and to other masses in the set of M_i. Such transmissions must be assumed to be reversible to preserve the condition of uniformity of T and p. The total effect of all individual transmissions will be $\Delta\psi'$, since inter-transmissions will cancel in the sum. Hence

$$\Delta\psi' = \sum_i (\Delta Q'_{t,i} - \Delta W'_{t,i}) = \sum_i \Delta Q'_{t,i} - \Delta W'_t. \tag{6.3.5}$$

But under reversible conditions

$$\sum_i \Delta Q'_{t,i} = \sum_i T\,dS'_i. \tag{6.3.6}$$

Using this, and (6.3.2) and (6.3.3),

$$dU' = T\sum_i dS'_i - \Delta W'_t. \tag{6.3.7}$$

Inserting this in equation (6.3.1) gives

$$\Delta W'_t = p\,dV' - \sum_i G_i\,dM_i. \tag{6.3.8}$$

Now we have already seen that in the spontaneous re-distribution $\sum_i G_i\,dM_i$ is negative.

Hence if a re-distribution of mass occurs spontaneously, the energy transmitted by work in a reversible change, i.e. with no frictional dissipation of any kind, is greater than $p\,dV'$ – i.e. is greater than the expansive work transmission done in the normal way by the whole mass. Thus the term $-\sum_i G_i\,dM_i$ gives effectively a change of a potential, and provided some device can be included to provide a work force against which this potential can operate, we shall be able to get useful output transmission of work from this potential.

In all real processes dissipation is present. We may represent it as a generalisation of the usual form and write, instead of (6.3.6),

$$\Delta Q'_t = T\sum_i dS'_i - \Delta W'_f,$$

whence

$$\Delta W'_t = p\,dV' - \sum_i G_i\,dM_i - \Delta W'_f. \tag{6.3.9}$$

Denoting the redistribution work transmission by $\Delta W'_{td}$ so that we may consider it apart from the work transmission of the total mass expansion, we have therefore

$$\Delta W'_{td} = -\sum_i G_i\,dM_i - \Delta W'_f. \tag{6.3.10}$$

If *no* device is present to obtain work transmission from the redistribution, $\Delta W'_{td} = 0$ and the whole of the potential change is dissipated, i.e.

$$-\sum_i G_i\,dM_i = \Delta W'_f, \tag{6.3.11}$$

and in these circumstances

$$\Delta Q'_t = T\sum_i dS'_i + \sum_i G_i\,dM_i. \tag{6.3.12}$$

Now the function G has another very interesting property, which can be seen as follows. Considering the potential for a mass M_i as G'_i we have

$$dG'_i = M_i\,dG_i + G_i\,dM_i. \tag{6.3.13}$$

But

$$dG_i = dU_i + p\,dV_i + V_i\,dp - T\,dS_i - S_i\,dT;$$

$$\therefore \quad dG_i = V_i\,dp - S_i\,dT. \tag{6.3.14}$$

Now if we stipulate not only that the pressure is uniform throughout M, but that in all the change which occurs both p and T remain constant, we shall then have

$$dG_i = 0; \tag{6.3.15}$$

$$\therefore \quad dG'_i = G_i\,dM_i. \tag{6.3.16}$$

Thus at constant pressure and temperature the potential G'_i has the unique characteristic that it is altered only by the addition of mass.

$$\therefore \quad G_i\,dM_i = dG'_i = dH'_i - T\,dS'_i$$

at constant temperature. Substitution in (6.3.12) gives

$$\Delta Q_t' = \sum_i dH_i' = dH'. \qquad (6.3.17)$$

This result occurs because we are considering that the pressure remains constant, as well as uniform, during the change, so that $dU' + p\,dV' = dH'$, and it is $\Delta W_{td}'$, i.e. the potential work, which we have assumed to be wholly dissipated, leaving the expansion work $p\,dV'$ unaffected.

Hence when a spontaneous re-distribution of mass occurs with no device present to extract the potential work transmission, the result will be that, if temperature and pressure are to be maintained constant, an input of energy by heating will occur equal in amount to the increase of enthalpy caused by the re-distribution. If the enthalpy after re-distribution at the controlled pressure and temperature is less than before, the result will be an *output* transmission of energy by heating.

In the general real case, where a device to extract the potential work is present, but dissipation of some of it occurs, we shall still have $\Delta Q_t'$ given from equation (6.3.9) by

$$\begin{aligned}\Delta Q_t' &= T\sum_i dS_i' - \Delta W_f \\ &= T\sum_i dS_i' - p\,dV' + \Delta W_t' + \sum_i G_i\,dM_i \\ &= T\sum_i dS_i' + \Delta W_{td}' + \sum_i dG_i'.\end{aligned}$$

$$\therefore \quad \Delta Q_t' = dH' + \Delta W_{td}'. \qquad (6.3.18)$$

The importance of the preceding discussion is to show that thermal effects will always accompany re-distribution of mass whether or not an effective device to extract the potential in work transmission is present. Hence this gives us a clue as to how we may detect that re-distribution is occurring, if no other evidence is available.

Now, of course, if because of confinement or some extent of insulation from environment the amount $\Delta Q_t'$ which includes $p\,dV'$ cannot be transmitted while maintaining pressure and temperature constant, these will tend to rise or fall, so that in any case we shall have indications of the results of re-distribution. If these indications are small, we may need very sensitive experiments to detect them, but in principle they must be present. And where the effects are large we may expect to detect them easily.

One further point must however be made. If G_i happens to be uniform throughout the mass M, re-distribution of M_i will not produce any $\Delta W'_{id}$ since the potential $-\sum_i G_i dM_i$ is then zero. Yet since a change of entropy occurs of amount $T dS'$ now identical with $\sum_i T dS'_i$, there will still be a thermal energy transmission $\Delta Q'_t = T dS'$ unaffected by re-distribution. How then can we distinguish between a thermal effect which arises from $T dS'$ without the existence of a potential, i.e. when G is uniform, and one which accompanies re-distribution when G is not initially uniform? The answer is an important step in our understanding. The distinction is that if G is uniform, no effect will arise *spontaneously*. We shall have deliberately to take some action on mass M in order to cause re-distribution and obtain the thermal effect, because equilibrium already exists. But if G is initially non-uniform, the thermal effects will occur spontaneously as the disequilibrium is reduced. Hence we have no difficulty in distinguishing between the two situations.

These points enable us now to proceed with the next major step in our development. The discussion so far depends upon the assumption that within a mass M at a uniform pressure and temperature we may nevertheless have portions in which the function $G = U + pV - TS$ is different. We have described the effects that will occur if this assumption is valid. Now we must see whether any phenomena in our experience can be considered to correspond to those effects and hence in what way, if any, the analysis can improve our understanding and control of such occurrences.†

6.4 Examples of mass re-distribution I. Phase change

Perhaps the most obvious case in our phenomenological experience where we can suppose that re-distribution of mass is taking place is in phase change. We know that a single pure substance such as water can exist as a gas, a liquid, or a solid, and if two of these phases are co-existing, we can see the possibility that the total mass may be distributed in

† The reader should, as an exercise, check that when a similar treatment to that done in this section for dU' is developed for dH', the result corresponding to equation (6.3.8) is

$$\Delta W'_d = -V' dp - \sum_i G_i dM_i. \qquad (6.3.19)$$

The term $-V' dp$ is the maximum work obtainable by a reversible *flow* process without re-distribution, just as in (6.3.8) $p dV'$ is the maximum work obtainable by a reversible *non-flow* process without re-distribution. In both cases the potential effect gives the same additional term, i.e. the possibility of re-distribution arising from *non-uniform* G is a work transmission additional to and quite distinct in principle from both expansion energy transmission and flow energy transmission.

varying proportions between the phases. In a single phase of a pure substance we already know that given p and T all other properties are fixed by the equation of state. Hence a single phase at uniform pressure and temperature can only have one value of G so that no potential phenomena will be exhibited. Thus we can only get energy transmission by work – other than the normal expansion or flow transmissions – from a single phase of a pure substance if p and/or T are not uniform.

When we consider two phases co-existing in saturation equilibrium, we have, taking first the example of liquid and vapour,

$$G_f = U_f + pV_f - TS_f, \tag{6.4.1}$$

$$G_g = U_g + pV_g - TS_g. \tag{6.4.2}$$

But

$$S_g - S_f = \frac{L}{T}, \tag{6.4.3}$$

where L is the latent heat, and

$$L = H_g - H_f. \tag{6.4.4}$$

Hence

$$G_g = G_f. \tag{6.4.5}$$

Thus the value of G is identical for two phases of a pure substance in equilibrium, and we shall have no potential effects. The points made in the previous section, however, will become more readily appreciated by considering this example further.

If a mass m_{fg} changes from the liquid to the vapour phase, $dS' = m_{fg} L/T$. We have of course a thermal effect, since an energy input by heating of amount $\Delta Q'_t = m_{fg} L$ is required, but no potential work output is obtainable. We have only the expansion work

$$p\,dV' = m_{fg}\,p(V_g - V_f).$$

The same arguments hold obviously for any two phases, solid/vapour or solid/liquid. The case of the three phases all together at the same pressure and temperature is rather more interesting. We have, for the solid phase,

$$G_s = U_s + pV_s - TS_s. \tag{6.4.6}$$

Now from any *two* of equations (6.4.1), (6.4.2) and (6.4.6) we can find a function of the form $\psi(p, T) = 0$ which will define a curve of pressure against temperature. Thus for example $\psi_{fg}(p, T) = 0$ will be a curve giving the locus of all p, T conditions for which $G_f = G_g$, i.e. it will define all conditions for which vapour and liquid are in equilibrium and a mass M can be divided between these two phases in any proportions for each p, T condition lying on $\psi_{fg}(p, T) = 0$. Similar curves $\psi_{gs}(p, T) = 0$ and $\psi_{sf}(p, T) = 0$ will exist for the solid/vapour and solid/liquid pairs. But the three curves can only meet at one single point, i.e. there can only be one single condition of pressure and temperature, which we may call p_3, T_3, at which the values G_f, G_g and G_s will all be equal. Hence only at one single pressure and temperature can we have a mass of a substance consisting of three parts M_f, M_g and M_s in the three different phases. The point p_3, T_3 is called the triple point.

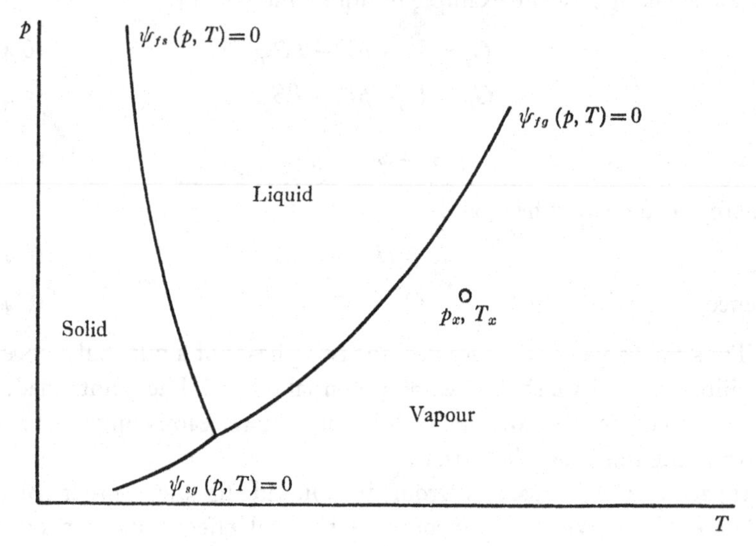

Fig. 6.4.1

If the pressure and temperature are different from p_3, T_3 and all three phases *are* initially present, one of them *must* spontaneously vanish, i.e. its mass must transfer to either or both of the other two phases. And if the pressure and temperature values do not lie on any of the three curves, only one phase can exist and if two are initially present one must spontaneously vanish by transfer of its mass to the other.

It should be noted that nothing we include here gives any information on the *rate* of such vanishing. Whether it occurs rapidly or slowly depends on other matters which are not our concern here. We are discussing only the direction in which changes will take place.

The diagram (Figure 6.4.1) illustrates the points made above.

If liquid and vapour are somehow simultaneously present at the conditions of p_x, T_x indicated there, then some liquid must spontaneously evaporate. This will of course require input thermal energy transmission to maintain the temperature constant and output work transmission to maintain the pressure constant, and if these do not occur sufficiently rapidly the temperature will tend to fall and the pressure tend to rise. The conditions will therefore move towards the ψ_{fg} curve, and this will continue until either only the vapour phase is left or a condition lying on the curve is reached. Similar descriptions apply for other conditions.

The behaviour described with reference to p_x, T_x is easily obtained in practice by heating up a liquid to temperature T_x under pressure higher than p_x and then injecting it into space containing superheated vapour. Mass is transferred from the liquid phase to the vapour phase, so that initially G_{fx} must be higher than G_{gx}. We have a potential difference so that in principle work additional to the expansion work could be obtained from such arrangement.

6.5 The existence of chemical species

We have seen from the previous section that if pressure and temperature are uniform throughout a mass of a pure substance, whether it is in one or more phases, G is also uniform, and no potential effects can arise.

Now from other aspects of our experience we know that we can have a mass of one gram at atmospheric pressure and temperature occupying a volume of 11 200 cm^3, or at the same temperature and pressure, occupying 700 cm^3. In the first case it is one gram of what we call hydrogen and in the second it is one gram of what we call oxygen. From other evidence we believe that it consists of more fundamental particles, protons, electrons, neutrons, etc., but from our thermodynamic point of view all that concerns us is that at the same temperature and pressure a unit of mass in these two cases has very different specific volume, and is clearly in two distinct 'states'. Thus we can anticipate that not only will V be different, at the same p and T, but so may U, S and G.

We know of course that for these particular cases our surmise is very quickly proved correct. For we do observe that when a mass begins by consisting of what we call hydrogen and oxygen, there is a noticeable spontaneous occurrence, after which at least some of the mass exhibits the characteristics of what we call water. Substantial transfer of energy is observed during the process, or rise of pressure and temperature if the transfer is not sufficient to maintain these constant. Thus all the characteristics of a potential arising from initial unequal values of G, with consequent re-distribution of mass to drive towards uniformity of G, are present. And all of this is occurring in a single phase. Since all molecular weights are different, all specific volumes in the gas phase at equal pressure and temperature must be different, and hence U, S and G may well be different, for different chemical substances. Thus when more than one such substance is present, we may expect G to be non-uniform, and all the phenomena of potential to be exhibited.

From this point of view the property G at a given temperature and pressure effectively determines thermodynamically the chemical species in which the unit of mass exists. It may be appropriate here to mention that it was J. W. Gibbs who first drew attention to the importance of the function $U+pV-TS$ as determining the potential, and related it to the chemical nature of a substance. It is for this reason that the symbol G is adopted for it, and it is usually named the chemical potential, or Gibbs function.

However, the facts we know about chemical behaviour include those of combination and decomposition in constant molecular, and hence mass, proportions. Thus in the example we discussed, the reaction is represented by the relation

$$2H_2 + O_2 \rightleftarrows 2H_2O. \tag{6.5.1}$$

From the atomic weights we know that the mass relations which occur are

$$2 \times 2 + 32 \rightleftarrows 2 \times 18. \tag{6.5.2}$$

Thus 1 kg of hydrogen will combine only with 8 kg of oxygen, no more, no less, to produce 9 kg of water.

Hence when chemical reaction occurs in the course of the mass re-distributions which we are studying, the re-distributions which can occur are subject not only to the general constraint $\sum_i M_i = \text{constant}$, but to the additional constraint imposed by the fixed proportions in which reaction must occur. As will be seen in Section 6.7, this has an important effect on the equilibrium values of G. But the matters discussed in previous sections still remain valid.

6.6 Examples of mass re-distribution II. Mixing of gases without reaction

Consider two different chemical species, K_α and K_β each in the gas phase at the same pressure and temperature, and initially separated from each other by a partition. For the present case we shall assume that K_α and K_β

have no reaction with each other at all, so that the masses M_α and M_β of K_α and K_β stay constant and no other chemical species is formed.

Now initially let M_α occupy a space A having initial volume V'_{a0} and M_β occupy an adjoining space B having initial volume V'_{b0}. We shall denote by M_a the mass in reference space A and by M_b the mass in reference space B. Initially $M_{a0} = M_\alpha$ and $M_{b0} = M_\beta$.

Now suppose that the partition is removed. We now have the situation that there is a total mass $M = M_\alpha + M_\beta$ in the space $V'_a + V'_b = V'$ at uniform pressure p and temperature T, but with initially $G_{a0} = G_{\alpha 0}$ in A and $G_{b0} = G_{\beta 0}$ in B. Since these are not equal there is a potential present. How will it operate?

The answer is of course that M_a and M_b will change, under the constraint

$$dM_a + dM_b = 0, \tag{6.6.1}$$

i.e. the gases inter-diffuse. The mass re-distribution which occurs is with reference to the space A and the space B.

The increment dM_a will consist in general of an increment dm_{ba} which has diffused from B to A, and of a decrement dm_{ab} which has moved from A to B,

$$dM_a = dm_{ba} - dm_{ab} = -dM_b. \tag{6.6.2}$$

Hence the total potential change is

$$\begin{aligned} dG' = dG'_a + dG'_b &= G_a\, dM_a + G_b\, dM_b \\ &= (G_a - G_b)(dm_{ba} - dm_{ab}). \end{aligned} \tag{6.6.3}$$

Since dG' must be negative, if not zero,

$$\left.\begin{aligned} dm_{ba} < dm_{ab} \quad &\text{if} \quad G_a > G_b; \\ dm_{ba} > dm_{ab} \quad &\text{if} \quad G_a < G_b. \end{aligned}\right\} \tag{6.6.4}$$

Thus more mass diffuses to B from A as long as $G_a > G_b$ than comes from B to A, and conversely if $G_a < G_b$. The opposite diffusions therefore become equal when $G_a = G_b$.

Now initially $G_{a0} = G_{\alpha 0}$ and $G_{b0} = G_{\beta 0}$. For G_a and G_b to become equal, each must change, and can only do so if the value for unit mass in a space is a function of the relative proportions of K_α and K_β in unit mass. It follows that when $G_a = G_b$ as functions of the proportions, the proportions of K_α and K_β in A and B respectively must be identical. Suppose that when G_a becomes equal to G_b a total mass m_α of K_α has moved from A to B and a total mass m_β of K_β has moved from B to A.

The relative proportions of K_α and K_β in A are then $(M_\alpha - m_\alpha)/m_\beta$ and in B are $m_\alpha/(M_\beta - m_\beta)$. Thus when $G_a = G_b$ we have

$$\frac{M_\alpha - m_\alpha}{m_\beta} = \frac{m_\alpha}{M_\beta - m_\beta}. \tag{6.6.5}$$

Hence by a well-known elementary theorem, each of these ratios is equal to the ratio of the sum of numerators to the sum of denominators, i.e. to M_α/M_β.

In short, the process only stops when the two different species K_α and K_β are uniformly spread out over the whole of V', so that the relative concentration everywhere is simply the proportions of the initial masses.

The associated effects of increase of entropy and of energy transmissions to maintain p and T constant will all in principle be present – although they may be difficult to detect if the potential difference between the gases is not large.

If now additionally we assume that the two gases are perfect, then we shall have

$$pV_\alpha = R_\alpha T, \quad pV_\beta = R_\beta T. \tag{6.6.6}$$

Now let N_α and N_β be respectively the number of moles of K_α and K_β in unit mass. Then

$$R_\alpha = N_\alpha R_m, \quad R_\beta = N_\beta R_m; \tag{6.6.7}$$

$$\therefore \quad \frac{V_\alpha}{N_\alpha} = \frac{V_\beta}{N_\beta} = V_{pg}, \tag{6.6.8}$$

where V_{pg} denotes the standard volume of one mole of perfect gas at the condition p, T.

Hence the initial volumes V'_{a0} and V'_{b0} are

$$\left.\begin{aligned} V'_{a0} &= M_\alpha V_\alpha = N_\alpha M_\alpha V_{pg}, \\ V'_{b0} &= M_\beta V_\beta = N_\beta M_\beta V_{pg}. \end{aligned}\right\} \tag{6.6.9}$$

The sum $V' = V'_a + V'_b$ will stay constant, since N_α, N_β are absolute constants for the species, V_{pg} is fixed by pressure and temperature, and K_α and K_β are assumed to be species which do not interact so that M_α and M_β are constant.

Thus if the process described occurs with perfect gases, there will be no volume change, i.e. $dV' = 0$.

Also since T is constant, there will be no internal energy change, since the internal energy of a perfect gas is a function of temperature only. Hence $dU' = 0$. Thus in the case of mixing of perfect gases, constant temperature and pressure conditions mean no transmission of energy.

It follows that in the mixing of two perfect gases, $dH' = 0$ and hence, from equations (6.3.17) and (6.3.18)

$$\Delta Q'_t = 0, \tag{6.6.10}$$

$$\Delta W'_{td} = 0. \tag{6.6.11}$$

The potential change $G_a dM_a + G_b dM_b$ is wholly dissipated and goes entirely to the increase of entropy since from equation (6.3.10)

$$\Delta W'_{td} = -[G_a dM_a + G_b dM_b] - \Delta W'_f = 0;$$

$$\Delta Q'_t = T[dS'_a + dS'_b] - \Delta W'_f = 0;$$

$$\therefore \quad T(dS'_a + dS'_b) = -[G_a dM_a + G_b dM_b]. \tag{6.6.12}$$

The foregoing discussion describes the process, the constraints and the terminal point. But since we can have no precise information on the values of dm_{ba} and dm_{ab}, and on the concentrations in these transfer elements during the process, it is not convenient for the calculation of the entropy or potential change.

To calculate these, having understood the *process* by using the concepts of the varying masses M_a and M_b it is more convenient to think of the constant masses M_α and M_β and to think of the entropy or potential change as due to expansion of each from the initial pressure to the partial pressure.

Let us first recall what is meant by partial pressure.

The pressure exerted by n_x moles of perfect gas occupying a volume X' is

$$p = \frac{n_x R_m T}{X'}. \tag{6.6.13}$$

Similarly if n_y moles of another perfect gas occupy a volume Y' the pressure is

$$p = \frac{n_y R_m T}{Y'}. \tag{6.6.14}$$

Now if the pressures in X' and Y' are equal, we have

$$p = \frac{n_x R_m T}{X'} = \frac{n_y R_m T}{Y'}. \tag{6.6.15}$$

And by the ratio theorem therefore, each is equal to

$$\frac{n_x R_m T + n_g R_m T}{X' + Y'}. \tag{6.6.16}$$

$$\therefore \quad p = \frac{n_x R_m T}{X' + Y'} + \frac{n_y R_m T}{X' + Y'}. \tag{6.6.17}$$

Thus the value of the equal pressures exerted by each gas in its own space is equal to the sum of two terms which are in fact the different pressures at which each gas would itself occupy the joint space. Hence when two perfect gases at initially equal pressures in separate spaces mix, we can finally regard each as occupying the joint space at a lower pressure, known as the partial pressure, the sum of the two partial pressures giving the total pressure.

Returning to our stated problem the constant masses M_α and M_β can each be regarded as expanding from p initially to

$$p\left(\frac{M_\alpha N_\alpha}{M_\alpha N_\alpha + M_\beta N_\beta}\right) \quad \text{and} \quad p\left(\frac{M_\beta N_\beta}{M_\alpha N_\alpha + M_\beta N_\beta}\right)$$

respectively.

We have the general theorem, for any mass that

$$dG' = V'\,dp - S'\,dT + G\,dM, \tag{6.6.18}$$

so that for a constant mass at constant temperature

$$dG' = V'\,dp. \tag{6.6.19}$$

For a perfect gas at constant temperature $V' = MRT/p$,

$$\therefore\ dG' = MRT\frac{dp}{p}. \tag{6.6.20}$$

Thus we have

$$dG'_\alpha = M_\alpha R_\alpha T\frac{dp_\alpha}{p_\alpha}, \tag{6.6.21}$$

and

$$dG'_\beta = M_\beta R_\beta T\frac{dp_\beta}{p_\beta}. \tag{6.6.22}$$

$$\therefore\ G'_\alpha = G'_{\alpha 0} + M_\alpha R_\alpha T\log(p_\alpha/p), \tag{6.6.23}$$

$$G'_\beta = G'_{\beta 0} + M_\beta R_\beta T\log(p_\beta/p). \tag{6.6.24}$$

Per unit mass we have

$$G_\alpha = G_{\alpha 0} + R_\alpha T\log(p_\alpha/p), \tag{6.6.25}$$

$$G_\beta = G_{\beta 0} + R_\beta T\log(p_\beta/p). \tag{6.6.26}$$

Since p_α and p_β are each less than p we note that the potential of *each* gas is reduced by mixing with the other.

The entropy change is of course the negative of the potential change, i.e.

$$S' - S'_0 = -[M_\alpha R_\alpha T\log(p_\alpha/p) + M_\beta R_\beta T\log(p_\beta/p)]. \tag{6.6.27}$$

At the end of the mixing process, G per unit mass of mixture is uniform throughout the mixture. In terms of the species present,

$$G = \frac{M_\alpha G_\alpha + M_\beta G_\beta}{M_\alpha + M_\beta},$$

which is of course uniform once the mixture is uniform, irrespective of the values of G_α and G_β in the mixture. Now it is obvious from equations (6.6.25) and (6.6.26) that in general G_α will not be equal to G_β. Equality can occur only for one particular mixture proportion.

Hence for any other proportion, the mixture will contain a non-uniformity of $G_\beta \neq G_\alpha$, although throughout the space the mean G is uniform. Thus there is an internal difference of potential still present. Something else has to happen. That something is *reaction*, or combination, which we discuss in the next section. There are in fact very few substances for which exact equality of potential can be obtained by mere mixing, and almost always some reaction occurs.

We have discussed the mixing process so far in terms of only two gases. However it is easy to visualise the extension to several gases M_α, M_β, M_γ, ..., etc., of K_α, K_β, K_γ, ..., etc. The same arguments indicate that, in the absence of reaction, mixing will occur until each substance is spread uniformly throughout the space giving concentrations

$$\frac{M_\alpha}{\sum_i M_i}, \quad \frac{M_\beta}{\sum_i M_i}, \quad \frac{M_\gamma}{\sum_i M_i}, \quad \ldots, \text{ etc.}$$

The uniform value of G throughout the mixture will be

$$\frac{\Sigma M_i G_i}{\Sigma M_i}$$

and the values G_α, G_β, G_γ, ..., G_i for the individual components in the mixture will *each* be less than the values $G_{\alpha 0}$, $G_{\beta 0}$, $G_{\gamma 0}$, ..., G_{i0} for the individual components as pure substances at the same temperature and total pressure. If the values G_α, G_β, ..., etc. in the mixture are *not* equal there is an internal potential which must lead to chemical reaction.†

† *Gibbs paradox.* It is frequent practice to derive the increase in entropy in the mixing process directly in terms of the partial pressure argument as we have done for the potential from equation (6.6.13) onwards. Then the question is raised – what about the removal of a partition separating two portions of the same gas at same temperature and pressure? The same arguments for the entropy case appear superficially plausible – yet we do not accept that there can in fact be an increase of entropy in such circumstances. This so-called 'paradox' is resolved by saying that the argument just does not apply because there is not in fact a continuous range of variation from one

6.7 Examples of mass re-distribution III. Introduction to chemical reaction

In the previous examples of mass redistribution each dM_i was independent of all the others, except for the constraint $\sum_i dM_i = 0$. This led to the conclusion that all G_i must be equal.

When a chemical combination or decomposition takes place it is known that it does so in constant proportions. Thus, as already noted in Section 6.5, one mass unit of hydrogen will react with eight mass units of oxygen, no more and no less. Hence in the reaction of hydrogen and oxygen to produce water we shall have

$$dM_{\text{oxygen}} = 8\,dM_{\text{hydrogen}}, \tag{6.7.1}$$

and

$$dM_{\text{water}} = -9\,dM_{\text{hydrogen}}. \tag{6.7.2}$$

Thus in mass units the reaction

$$2H_2 + O_2 \rightleftarrows 2H_2O \tag{6.7.3}$$

can be expressed as, since mass is conserved,

$$dM_{\text{hydrogen}}[1+8-9] = 0. \tag{6.7.4}$$

Hence the constraint $\sum_i dM_i = 0$ to which all our discussions are subject is modified in a particular way when any of the species K_i can react with any other. If all species present may interact we shall have a set of mass ratio coefficients obtained from the reaction equation in the same way as those in (6.7.4) are obtained from (6.7.3). Thus instead of

$$dM_\alpha + dM_\beta + dM_\gamma + \ldots + dM_i + \ldots + dM_z = 0, \tag{6.7.5}$$

it will be possible, and indeed necessary, to write

$$dM_i[a_1 + a_2 + \ldots + a_i + \ldots + a_r - b_1 - b_2 \ldots - b_k \ldots - b_s] = 0, \tag{6.7.6}$$

i.e. the constraint

$$\sum_i dM_i = 0 \tag{6.7.7}$$

becomes

$$\sum_j a_j - \sum_k b_k = 0. \tag{6.7.8}$$

substance to another so that it is not correct to regard the case of two portions of identical gas as a limit of two 'nearly identical' gases.

The present treatment which begins by discussing the nature of the process in terms of the redistribution of mass shows that there is no paradoxical element at all in the situation. If the two portions are identical gas, then G will be uniform throughout originally, and there will automatically, from equation (6.6.3), be no change in entropy. We use the partial pressure argument for convenience in calculating the actual change when G is not initially uniform.

The mass ratio coefficients are classed into two sets a_j and b_k, so that all will be positive, the b_k being those which, from the reaction equation, would all have negative signs. In the reaction equation itself these would correspond to terms which would be regarded as *products* of the reaction. Naturally, since the reaction may go either way, such *naming* of the two classes as 'reactants' and 'products' is arbitrary. But it is a necessary convention for thinking about the reaction process.

It is these relations which must now be substituted into the equilibrium condition

$$\sum_i G_i dM_i = 0. \tag{6.7.9}$$

We have and

$$\left.\begin{aligned} dM_j &= a_j\, dM_i \quad \text{for all } j, \\ dM_k &= -b_k\, dM_i \quad \text{for all } k. \end{aligned}\right\} \tag{6.7.10}$$

Either the set j or the set k may include the substance K_i which is selected as the mass base in equation (6.7.6), i.e. we may select either one of the reactants or one of the products.† In either case a_i or b_i will be unity, but can be included in the general terms without noting this fact until the calculation stage.

Using relations (6.7.10) the equilibrium condition (6.7.9) becomes

$$\sum_j a_j G_j = \sum_k b_k G_k. \tag{6.7.11}$$

And using (6.7.8) also, we have therefore

$$\frac{\sum_j a_j G_j}{\sum_j a_j} = \frac{\sum_k b_k G_k}{\sum_k b_k}. \tag{6.7.12}$$

Equation (6.7.11) *or* (6.7.12) gives the equilibrium condition for substances which may react chemically. Equation (6.7.12) has a most interesting interpretation. We recall from Section 6.6 that in the gas mixture of all 'reactants' and all 'products' the individual values of each G_j and G_k will be affected by the proportions of *all* the others, as well as by its own value as a pure substance. Hence the *actual* mass proportions of *all* substances present are involved in determining each G_j and G_k in equation (6.7.12). The coefficients a_j and b_k on the other hand do not include the actual mass proportions present at all, but only the proportions of the reaction relations. Equation (6.7.12) shows that equilibrium does *not* require that all individual G_i per unit mass of component K_i be

† In equations (6.7.10) it is assumed that K_i is one of the reactants, hence the negative sign for dM_k.

equal, *nor* that the mean $\bar{G}_j$ per unit mass of reactants, which would be $\sum_j M_j G_j / \sum_j M_j$ be equal to the mean $\bar{G}_k$ per unit mass of products, $\sum_k M_k G_k / \sum_k M_k$. It shows instead that, given the actual G_j and G_k for each individual substance determined by all the actual M_j and M_k and the total pressure and temperature, the equilibrium is determined when a mean for the reactants in reaction (stoichiometric) proportions is equal to a corresponding mean for the products in their stoichiometric proportions.

Thus, for example, in the case of hydrogen, oxygen, and water, we have the reaction equation

$$H_2 + \tfrac{1}{2}O_2 \rightleftharpoons H_2O,$$

which indicates the stoichiometric reaction proportions in mass units of 8 of oxygen and 9 of water to one of hydrogen. Regarding hydrogen and oxygen as reactants and water as the sole product, we have

$$a_{H_2} = 1, \quad a_{O_2} = 8, \quad b_{H_2O} = 9.$$

Now suppose that at some particular temperature and pressure, hydrogen and oxygen and water are all three present together in proportions such that $M_{O_2}/M_{H_2} = 2$ and $M_{H_2O}/M_{H_2} = 100$. These proportions will, together with the temperature and pressure, determine the actual value of G_{H_2}, G_{O_2} and G_{H_2O} in the mixture, but this mixture will only be in equilibrium if

$$G_{H_2} + 8G_{O_2} = 9G_{H_2O}, \tag{6.7.13}$$

i.e. per unit mass of this stoichiometric *weighting* – we cannot call it a stoichiometric mixture, since the mixture proportions are very different – we have

$$\frac{G_{H_2} + 8G_{O_2}}{9} = G_{H_2O}. \tag{6.7.14}$$

If the left-hand side is greater than the right, more combination must occur. If the right hand side is the greater, more decomposition will take place.

The behaviour is as if, whatever the actual proportions present, which determine the actual G values, we have to consider the 'stuff' of unit mass of water and compare its Gibbs function value as the compound, with a mean Gibbs function for its constituents in the formative proportions for water. It is a comparison between two possible 'states' of a unit of mass, the 'state' of being all water, as against the 'state' of being eight-ninths oxygen and one-ninth hydrogen, under general conditions set by pressure, temperature, and the actual amounts of each substance present.

For perfect gases we recall that, as in equations (6.6.25) and (6.6.26), the actual value of G for any of the constituent substances in the mixture is given by

$$G_i = G_{i0} + R_i T \log(p_i/p). \tag{6.7.15}$$

Equations (6.7.11) and (6.7.15) are the foundation equations upon which all procedure of calculating and controlling chemical equilibrium is based. Taken together they lend to an equation for the so-called *equilibrium 'constant'*. The development of this concept and the methods for its use are more appropriately studied in specialist texts on chemical thermodynamics or chemical engineering.

We conclude our own discussion by drawing attention to some subsidiary relations.

Mathematically we may express the total Gibbs function for the whole mass $M = \Sigma M_i$ as

$$G' = G'(p, T, M_\alpha, M_\beta, \ldots, M_i, \ldots). \tag{6.7.16}$$

$$\therefore \quad dG' = \frac{\partial G'}{\partial p}dp + \frac{\partial G'}{\partial T}dT + \sum_i \frac{\partial G'}{\partial M_i}dM_i$$

$$= V'\,dp - S'\,dT + \sum_i \frac{\partial G'}{\partial M_i}dM_i. \tag{6.7.17}$$

But for any one of the masses we also have

$$G'_i = M_i G_i;$$

$$\therefore \quad G' = \sum_i M_i G_i = \sum_i G'_i; \tag{6.7.18}$$

$$\therefore \quad dG' = \sum_i dG'_i. \tag{6.7.19}$$

Also we have

$$dG'_i = V'_i dp_i - S'_i dT_i + G_i dM_i. \tag{6.7.20}$$

Now if we consider each mass M_i as exerting its partial pressure in the manner of a perfect gas, V'_i will be the same for all M_i since it will be the total volume V' and p_i will be the partial pressure. We assume the temperature to be uniform. Hence, using (6.7.20) in (6.7.19) we have

$$dG' = V'\sum_i dp_i - dT\sum_i S'_i + \sum_i G_i dM_i. \tag{6.7.21}$$

But

$$p = \sum_i p_i,$$

$$\therefore \quad dp = \sum_i dp_i,$$

and

$$\sum_i S'_i = S'.$$

Hence comparing coefficients in equation (6.7.17) and (6.7.21) we find that

$$G_i = \frac{\partial G'}{\partial M_i}. \tag{6.7.22}$$

Thus the individual G values for unit mass of any one substance in the mixture is the partial differential of the total potential of the whole mass of mixture with respect to mass of that one substance, the other masses being assumed to remain unaltered. This notation of partial differentials will often be found instead of reference to individual G_i.

Another aspect to which attention should be drawn is the frequent use of molar quantities and ratios instead of mass quantities and ratio. Thus instead of referring to G' as the potential for a total mass M we may refer to μ' as the potential for a total number of moles ν. The potential per mole is then denoted by μ, which corresponds to G per unit mass. The potential per mole of a given constituent is then μ_i and is used instead of G_i and we have also

$$\mu_i = \frac{\partial \mu'}{\partial \nu_i}, \tag{6.7.23}$$

where ν_i denotes the number of moles of species K_i present in the mixture. The use of molar quantities has the advantage that the partial pressures, for perfect gases, are proportional directly to the number of moles.

Recalling that N_i is the number of moles in unit mass of substance K_i we have the relationships

$$\nu_i = N_i M_i; \tag{6.7.24}$$

$$\mu_i = \frac{G_i}{N_i}, \quad G_i = N_i \mu_i; \tag{6.7.25}$$

$$G' = \sum_i M_i G_i = \sum \frac{\nu_i}{N_i} G_i = \sum \nu_i \mu_i = \mu'. \tag{6.7.26}$$

In the chemical reaction equations, the mass ratio coefficients a_j and b_k will have corresponding molar ratio coefficients which we shall denote by α_j and β_k where

$$\alpha_j = a_j N_j, \quad \beta_k = b_k N_k. \tag{6.7.27}$$

Hence

$$a_j G_j = \frac{\alpha_j}{N_j} N_j \mu_j = \alpha_j \mu_j. \tag{6.7.28}$$

Thus the general chemical equilibrium condition (6.7.11) looks exactly similar when transferred to molar terms, viz.

$$\sum_j \alpha_j \mu_j = \sum_k \beta_k \mu_k. \tag{6.7.29}$$

But the relation (6.7.12) becomes

$$\frac{\sum_j \alpha_j \mu_j}{\sum_j \alpha_j/N_j} = \frac{\sum_k \beta_k \mu_k}{\sum_k \beta_k/N_k}. \tag{6.7.30}$$

Equation (6.7.30) has no simple interpretation in molar stoichiometric terms such as we could give to (6.7.12) in mass stoichiometric terms. However, the important equilibrium condition is adequately given by (6.7.29) in molar terms, which corresponds to (6.7.11). We have already noted that all further development depends upon combining equation (6.7.11) with (6.7.15).

Now in molar terms, equation (6.7.15) becomes

$$\mu_i = \mu_{i0} + \frac{R_i T}{N_i} \log (p_i/p). \tag{6.7.31}$$

$$\therefore \quad \mu_i = \mu_{i0} + R_m T \log (p_i/p); \tag{6.7.32}$$

$$\therefore \quad \mu_i = \mu_{i0} + R_m T \log \left(\frac{\nu_i}{\sum_i \nu_i}\right). \tag{6.7.33}$$

The convenience of having the universal perfect gas constant in this equation as well as the direct relationship between partial pressure and number of species moles far outweighs other considerations. Hence all treatment of chemical thermodynamics uses molar quantities, ratios and relationships for discussion and evaluation.

7

SURVEY

– the golden glisten of the big trout in the roughwater, the landing net, quick! another capture before the end of the drift – but here is the end of the drift.

Neil M. Gunn

With the conclusion of the preceding chapter we have come to the end of the immediate task of the present study, for we have now answered the question posed in Chapter 5. We see now, from Section 6.7 of Chapter 6, that substances which are chemically different are so because they have different values of potential, Gibbs function – or free energy as it is sometimes called – per unit mass, even at the same temperature and pressure. In short – as a layman might justly say was obvious from the start – if we have different substances we do not necessarily have uniformity even if we have a uniform temperature and pressure environment. We now know, however, something which is not at all obvious to the layman, that an effective 'uniformity', or rather equilibrium, can exist even if distinguishable chemical species are present. For our purpose, however, we are more interested in the fact that provided we do not have the necessary conditions for equilibrium the process of chemical reaction will occur. As one example of mass re-distribution with different initial G values this process is then bound to be accompanied by thermal effects as described in Section 6.3 of Chapter 6. It is this which gives us the answer we sought in Chapter 5, i.e. the supply of energy at a higher temperature. For the value of $\Delta Q_t'$ given by equation (6.3.17) will, if dH' is negative, be an output transmission of energy and unless this can escape rapidly enough to the surrounding environment the temperature is bound to rise. With this rise the energy can be transmitted, to provide work by the processes outlined in the earlier chapters.

What we call a fuel, or rather a combustible mixture, is a mixture of components with different initial values of G when separate, such that the sum of the enthalpies of the mixture at *equilibrium* proportions at standard pressure and temperature is *less* than the enthalpies of the original separate components at the same standard pressure and temperature. It is this fall in enthalpy which gives what we call the emission of 'heat', i.e. $\Delta Q_t' = dH'$ with dH' negative. And it is by the use of transmitted energy of this type that modern industrial power has operated ever since its inception some 200 years ago. In doing so it has been limited not only incidentally by the suitability of materials of construction, but inherently by the theorems regarding convertibility established in our earlier chapters.

But our study also reveals other possibilities – if we look at equations (6.3.18) and (6.3.10) instead of (6.3.17), and at the related discussion. We observe then that if we could introduce into the reaction process some device which could provide a work force against which the chemical potential difference could operate, we could get energy transmission by work directly, without the intermediary of transmission by heating – and hence without the inherent limitation of conversion ratio dependent on temperature. Is there a real possibility of this – or shall we again be faced with the situation described in Section 2.7 of Chapter 2, where we could not take advantage of non-uniformities smaller in scale than the dimensions of our working device? Now it happens that we *can* in some cases in principle overcome the scale limitation – not because we can devise a Maxwell 'demon' small enough to separate out slow and fast molecules – but because the energy situation which appears thermodynamically as chemical potential is related to electric potential of atoms and their accompanying electrons. Hence provided we can set up the components of combustible mixture in such a way that electrons or ions can move under the action of these potentials we achieve effectively the kind of device which can extract the $\Delta W_{td}'$ of equations (6.3.10) and (6.3.18). Such a device is called a fuel cell and much research work is being directed towards the realisation of its possibilities. We note of course that no such device can ever achieve 100% conversion of the initial chemical potential into work because, as shown by equation (6.3.10), there will always be some internal dissipation. Nevertheless the important point is that such a direct utilisation concept is free from the rejection ratio and conversion ratio requirements of the usual process of a conversion cycle.

Electrical cell devices such as the common acid battery are other useful examples in which chemical potential difference can be utilised directly in the performance of work $\Delta W_{td}'$. Again the phenomenon of internal cell dissipation is present. More important, however, the manufacture of these chemicals requires more input energy than is obtained in output, so that they do not provide a solution to the problem of producing work.

Some mention should also be made in this concluding survey of electrical, magnetic, and capillary effects, all of which we know of, and yet none of which has received any discussion in this book. We shall briefly indicate how they would in fact enter in a more comprehensive treatment of thermodynamics. This can be done most conveniently by referring back to Chapter 1. The initial stage where we began with the Newtonian equation of motion was completely general, i.e. the equation

$$md(\tfrac{1}{2}v^2) = \sum_r F_{s,r}\,ds$$

can include forces of any origin, including electric, magnetic, and capillary forces. Subsequently, however, when we changed to the discussion of a fluid element, we tacitly omitted all kinds of stress in the material other than the fluid pressure p. Regarding this as a generalised force concept, we have conjugate to it the generalised displacement dV, so that the term $p\,dV$ enters in the equation for ΔW_t.

Now if electric, magnetic, and capillary forces are also present in the material, these can be included in a similar manner, i.e. we can say that to any specific type of force F_k there will correspond some conjugate displacement dX_k such that a term $F_k\,dX_k$ enters into the expression for ΔW_t, the term $p\,dV$ being only one example. In the case of electric field as the force, the conjugate displacement is polarisation, for magnetic intensity we have magnetic moment as conjugate, and to surface tension the conjugate displacement is surface area. When these ideas are included the equation corresponding to (1.4.3) becomes

$$\Delta W_t = \sum_k F_k\,dX_k - d(\tfrac{1}{2}v^2) - \sum_k d\Phi_k - \Delta W_f, \tag{7.1}$$

where we include now also the possibility of other potential fields due to other bodies, and where ΔW_f includes *all* dissipative effects.

The arguments regarding conservation of energy as in Chapter 1, Section 1.6 proceed as before and we arrive at the conclusion

$$\Delta Q_t = dU + \sum_k F_k\,dX_k - \Delta W_f. \tag{7.2}$$

Equation (7.2) is the basis for the thermodynamic treatment of all other physical effects known to occur. Not only can it comprehend those mentioned above, as well as the simple case of fluid pressure, but the various stresses which occur in solids are also amenable to this treatment. By such means specialist texts can develop a full account of the thermodynamics of material properties, and of processes involving these properties.

Finally we should also include some reference to nuclear energy in this concluding survey. It takes only a little imagination to visualise that essentially the same arguments as were developed in Chapter 6 are also applicable. The fundamental concept is that of re-distribution of mass between different energy states. In the case of nuclear reactions we have the additional complication that we no longer can treat conservation of mass and conservation of energy as separately valid but must allow for mass changing also under an overall conservation of energy plus mass. At present the utilisation of nuclear energy is still based on the transmission of energy by heating – a nuclear equivalent to the $\Delta Q'_t = dH'$ process – and the resultant operation of a work-producing cycle. But again some research activity is directed towards the possibilities of finding some nuclear equivalent to the fuel cell.

APPENDIX. GENERAL ANALYSIS OF COMPRESSION AND EXPANSION PROCESSES

From now on man willy-nilly finds his own image stamped on all he looks at.

Pierre Teilhard de Chardin

Introduction

The preceding text sets out the essentials of the conceptual scheme of thermodynamics which has been developed. The uses of this scheme are many, and a course in thermodynamics normally continues to develop various techniques to exploit the conceptual scheme for different purposes. In this Appendix we give only one example of such subsequent development, the methods used for the analysis of machines performing compression and expansion duties on a pure substance.

A. 1 Mechanical efficiency of expansion and compression processes

We now make practical use of the property entropy by using it in the equations already established for any real process change. Substituting

$$T\,dS = dU + p\,dV \tag{A. 1.1}$$

into (1.6.6) gives

$$\Delta Q_t = T\,dS - \Delta W_f, \tag{A. 1.2}$$

and into (2.2.7) and (2.2.3) gives

$$\Delta W_d = -V\,dp + \Delta Q_t - T\,dS - d(\tfrac{1}{2}v^2) - d\Phi \quad \text{(flow process)} \tag{A. 1.3}$$

$$\Delta W_t = p\,dV + \Delta Q_t - T\,dS \quad \text{(non-flow process).} \tag{A. 1.4}$$

Case 1. Flow expansion process

Let us consider first a flow process occurring under conditions where the change of potential energy can be neglected – as, for example, in horizontal flow – and also where change of bulk kinetic energy $\frac{1}{2}v^2$ can be neglected. We then have

$$\Delta W_d = -V\,dp + \Delta Q_t - T\,dS$$
$$= -V\,dp - \Delta W_f. \tag{A. 1.5}$$

If dp is negative, i.e. an expansion process, work transmission from the reference substance will be positive, and will be a maximum possible value *for any given $V\,dp$* if there is no friction. Let us denote the actual thermodynamic path element (recalling that quasi-equilibrium properties are implied throughout) by the suffix symbol a. Then the maximum transmitted work from the *actual* path would be $-V_a dp_a$ but the *actual* transmitted work from this path will be less, i.e. $-V_a dp_a - \Delta W_f$.

Thus we have

$$\left.\begin{aligned}\Delta W_{dm} &= -V_a dp_a,\\ \Delta W_{da} &= -V_a dp_a - \Delta W_f.\end{aligned}\right\} \qquad \text{(A 1.6)}$$

The actual frictional energy dissipation is $\Delta W_f = \Delta W_{dm} - \Delta W_{da}$ and we define the *mechanical efficiency* of the process as

$$\eta_{ae} = \frac{\Delta W_{da}}{\Delta W_{dm}} = \frac{-V_a dp_a - \Delta W_f}{-V_a dp_a}. \qquad \text{(A. 1.7)}$$

From (A. 1.5) we have also

$$\eta_{ae} = \frac{-V_a dp_a + \Delta Q_{ta} - T_a dS_a}{-V_a dp_a}. \qquad \text{(A. 1.8)}$$

Now since $\quad dH = dU + p\,dV + V\,dp = T\,dS + V\,dp, \qquad$ (A. 1.9)

we also have $\quad \eta_{ae} = \dfrac{-dH_a + \Delta Q_{ta}}{-dH_a + T_a dS_a} = \dfrac{\Delta Q_{ta} - dH_a}{T_a dS_a - dH_a}. \qquad$ (A. 1.10)

The formula of equation (A. 1.10) is quite general and can be used for any substance undergoing any flow expansion process. Many practical flow expansion processes are carried out adiabatically, so that $\Delta Q_{ta} = 0$. Then we have

$$\eta_{ae} = \frac{-dH_a}{T_a dS_a - dH_a} \quad \text{(adiabatic).} \qquad \text{(A. 1.11)}$$

These relations express the actual mechanical efficiency at any point in the process path. A mean overall efficiency can be calculated from observed results as

$$\bar{\eta}_{ae} = \frac{\int \Delta Q_{ta} - \int dH_a}{\int T_a dS_a - \int dH_a} = \frac{Q_t - (H_2 - H_1)}{\int_1^2 T_a dS_a - (H_2 - H_1)}. \qquad \text{(A. 1.12)}$$

Case 2. Non-flow expansion process

For this case, such as a gas expanding in a cylinder, we have, from (A. 1.4),

$$\begin{aligned}\Delta W_t &= p\,dV + \Delta Q_t - T\,dS\\ &= p\,dV - \Delta W_f.\end{aligned} \qquad \text{(A. 1.13)}$$

Arguments similar to the foregoing apply and we have again

$$\eta_{ae} = \frac{p_a\, dV_a - \Delta W_f}{p_a\, dV_a} = \frac{p_a\, dV_a + \Delta Q_t - T_a\, dS_a}{p_a\, dV_a}. \quad \text{(A. 1.14)}$$

But

$$p_a\, dV_a - T_a\, dS_a = -dU_a,$$

$$\therefore \quad \eta_{ae} = \frac{\Delta Q_t - dU_a}{T_a\, dS_a - dU_a}. \quad \text{(A. 1.15)}$$

For the adiabatic case

$$\eta_{ae} = \frac{-dU_a}{T_a\, dS_a - dU_a}, \quad \text{(A. 1.16)}$$

and for the general case the average for a complete expansion is

$$\bar{\eta}_{ae} = \frac{Q_t - (U_2 - U_1)}{\int_1^2 T_a\, dS_a - (U_2 - U_1)}. \quad \text{(A. 1.17)}$$

Case 3. Compression processes (flow and non-flow)

For compression processes the basis equations (A. 1.5) and (A. 1.13) are the same as before. But now dp is positive or dV is negative, and the presence of friction means that *more* work has to be transmitted than if no friction were present. Hence in this case mechanical efficiency is defined as the reciprocal of the previous definition, i.e. it is now the ratio of the work which would have been required in the absence of friction for the actual compression path, to the actual work required. Hence we have, corresponding respectively to (A. 1.10), (A. 1.11), (A. 1.12) and (A. 1.15), (A. 1.16), (A. 1.17) the results

Flow processes

$$\eta_{ac} = \frac{dH_a - T_a\, dS_a}{dH_a - \Delta Q_{ta}}; \quad \text{(A. 1.18)}$$

adiabatic

$$\eta_{ac} = \frac{dH_a - T_a\, dS_a}{dH_a}; \quad \text{(A. 1.19)}$$

$$\bar{\eta}_{ac} = \frac{(H_2 - H_1) - \int_1^2 T_a\, dS_a}{(H_2 - H_1) - Q_t}. \quad \text{(A. 1.20)}$$

Non-flow processes

$$\eta_{ac} = \frac{dU_a - T_a\, dS_a}{dU_a - \Delta Q_t}; \quad \text{(A. 1.21)}$$

adiabatic $$\eta_{ac} = \frac{dU_a - T_a\,dS_a}{dU_a};\qquad \text{(A. 1.22)}$$

$$\bar{\eta}_{ac} = \frac{(U_2-U_1)-\int_1^2 T_a dS_a}{(U_2-U_1)-Q_t}.\qquad \text{(A. 1.23)}$$

Since in the absence of friction $\Delta Q_t = T\,dS$ each formula gives automatically $\eta = 1$ for the corresponding frictionless case.

Clearly the efficiency is calculable from these formulae provided we can establish values of enthalpy change, internal energy change, and values of the integral $\int T\,dS$.

A. 2 Isentropic efficiency and path efficiency

Since many expansion and compression processes are carried out in practice under adiabatic conditions, interest often centres on equations (A. 1.11), (A. 1.16), (A. 1.19), (A. 1.22), and the corresponding averages. It has also been customary to define an entirely different concept of mechanical efficiency for these cases. This is called the isentropic efficiency and its conceptualisation is as follows. (Our discussion refers for convenience only to the flow expansion case, but the reader can easily make the required conversion to the other cases.)

If friction were absent the dissipation would also be absent, and hence the thermodynamic path followed would be different from the actual path. In the particular case of an adiabatic change, the ideal frictionless path would be an isentropic one. Hence for the adiabatic case, another possible concept of the maximum amount of work obtainable is $-V_s\,dp_s$ where the suffix s denotes constant entropy. Hence we define isentropic efficiency in flow expansion as

$$\eta_{se} = \frac{\text{actual work}}{-V_s\,dp_s} = \frac{-dH_a}{-dH_s}.\qquad \text{(A. 2.1)}$$

Similar relations apply to the other processes.

For reference purposes it is desirable to have an accepted name to distinguish η_a from η_s and the name we adopt is *path efficiency*, to indicate that it is based on the optimum possibilities of the actual (quasi-equilibrium) thermodynamic path. There are two obvious advantages of the path efficiency as compared with the isentropic efficiency. The first is that it is applicable to *any* process, whether adiabatic or not, whereas the isentropic efficiency is meaningless for a process during which heating or cooling occurs. Most practical non-flow compression or expansion processes, e.g. reciprocating air compressors and the like, are not in fact adiabatic, and the use of isentropic efficiencies is misleading. The second is that even for adiabatic processes, since η_a is based on the actual path, and is in fact the ratio of $(-V\,dp_a - \Delta W_f)/(-V_a\,dp_a)$, the difference between numerator and denominator *is* a measure of the actual losses from which actual friction forces can be obtained. The isentropic efficiency cannot give a correct estimate of the losses. In the expansion process it underestimates them.

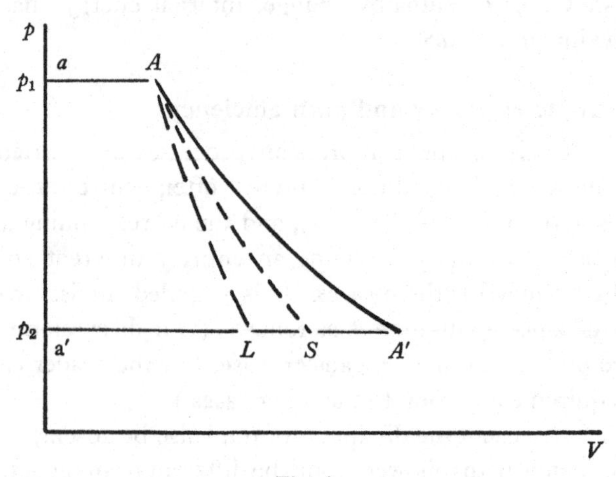

Fig. A.2.1

The difference between the two measures of mechanical efficiency for adiabatic processes can be illustrated by reference to Figure A. 2.1.

The line AA' represents the actual thermodynamic path, and the area $AA'L$† represents the actual loss $\int \Delta W_f$. The actual work done is represented in magnitude by the area $aALa'$ and

$$\eta_a = \frac{aALa'}{aAA'a'}. \tag{A. 2.2}$$

The line AS represents the ideal expansion which would occur under isentropic conditions if there were no friction, and

$$\eta_s = \frac{aALa'}{aASa'}. \tag{A. 2.3}$$

We see that necessarily $\eta_a < \eta_s$.

For a compression process the corresponding figure is shown in Figure A. 2.2.

We have
$$\eta_{ac} = \frac{a'A'Aa}{a'LAa}, \tag{A. 2.4}$$
and
$$\eta_{sc} = \frac{a'SAa}{a'LAa}. \tag{A. 2.5}$$

Here $\eta_{ac} > \eta_{sc}$ and the isentropic efficiency overestimates the actual friction.†

However, there is one other important aspect which should be borne in mind. Naturally in the expansion case the numerator for both isentropic and path efficiencies is the same, since it is the actual work transmission obtained. The difference is in the denominator. Now the denominator in the isentropic case is the work which would be obtained

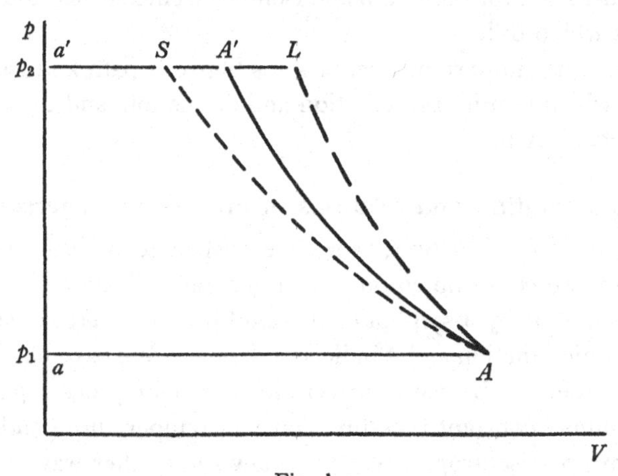

Fig. A.2.2

in an ideal frictionless process, hence it does represent an *ideal* maximum possible amount of work. In the path efficiency case the denominator is the work potential – i.e. $\int V_a\, dp_a$ – along the actual path, and in this the temperature is higher than it would be in the frictionless case, because of friction. Thus the denominator is a larger number in this case than without friction – but this larger number is only obtained because of friction and does not correspond to any possible amount of actual work transmission. As we have seen the *actual* transmission is

$$-\int V_a\, dp_a - \int T_a\, dS_a.$$

† N.B. *It is important to note* that while in Figures (A 2.1) and (A 2.2) the lines AA' and AS represent thermodynamic paths, i.e. successive states of the substance, the line AL does not. Its only significance is to indicate $\int \Delta W_f$ by the area $AA'L$.

Thus the isentropic efficiency has the advantage of indicating the ratio of the actual work obtained in a real adiabatic process to the maximum possible under ideal frictionless adiabatic conditions. This is obviously a matter of some interest and significance, in that it compares actual realisation with an ideal best. The interest of the path efficiency is just as great, and its significance is just as important, but quite different. It compares actual realisation with the total of actual realisation plus actual losses. There is considerable utility in practice for both concepts of mechanical efficiency.

The reader should argue through the comparable situation for adiabatic compression where this time the *denominators* – i.e. the actual work input – are the same for both path and isentropic efficiencies, and it is the numerators which differ.

There are other important relationships between path and isentropic efficiencies which require presentation and discussion, and these will be found in Section A. 6.

A. 3 Representation and relations of process characteristics

In any discussion up to now, when we wished to indicate a general process path, we could do so only by a general indication of a line on a state diagram, or by a symbol of the kind $[a\int dY$ where Y was some thermodynamic function and a indicated the path. We have nevertheless occasionally referred by name to certain particular process paths, i.e. constant volume, constant pressure, constant temperature (isothermal), constant entropy (isentropic). We seek now some other way of characterising a process path more generally.

Consider the diagram of Figure A. 3.1.

The element aa' of the process path may be defined completely by its co-ordinates p, V and by its slope dp/dV. These can all be lumped together to give one comprehensive specification of the element by considering the quantity

$$\frac{dp/p}{dV/V} = \frac{\text{fractional change of pressure}}{\text{fractional change of volume}}. \tag{A. 3.1}$$

We may denote this ratio of fractional change by the symbol $-n_a$ using the negative sign since most frequently we shall expect pressure increase when volume diminishes. Thus at the point a on the path we have

$$\frac{V}{p}\frac{dp}{dV} = -n_a. \tag{A. 3.2}$$

Now in the most general case n_a will differ along the path, and we shall be no more advanced. But we may nevertheless define a particular *type* of process, as that for which the value of the fractional change ratio of pressure to volume is constant, i.e.

$$\frac{V}{p}\frac{dp}{dV} = -n, \tag{A. 3.3}$$

all along the process path. Such a type of process is called *polytropic*, and is much more general than any of the four special processes we have discussed previously. It includes each of these as special cases, as will shortly be evident.

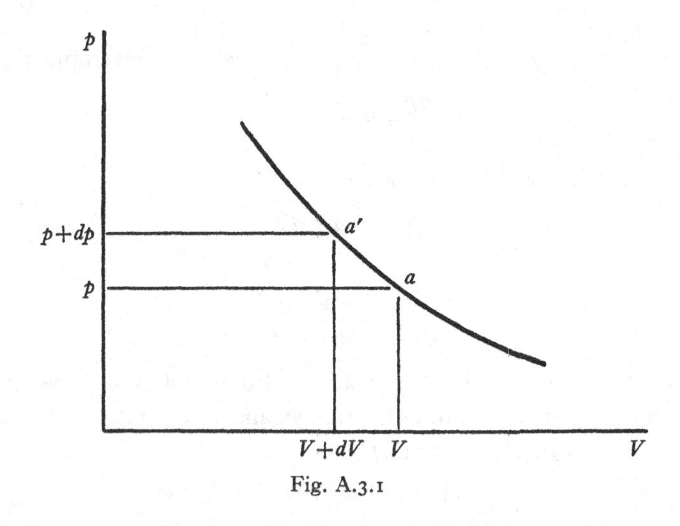

Fig. A.3.1

The defining polytropic equation (A. 3.3) can be written alternatively as

$$\frac{dp}{p} = -n\frac{dV}{V}, \tag{A. 3.4}$$

and therefore integrates immediately to

$$pV^n = \text{constant}. \tag{A. 3.5}$$

Thus, we see that for any substance, we have the following particular values of n corresponding to two of the frequently discussed particular process paths:

$n = 0$ constant pressure process;

$n = \infty$ constant volume process.

It is also evident that for a perfect gas $n = 1$ is the case of an isothermal process, since then pV is constant.

Hence the use of different values of n will allow us to range widely over different processes, while still allowing analytic treatment by classifying them all as of the same polytropic type defined by equation (A. 3.4) or (A. 3.5).

A. 4 Polytropic process characteristics for a perfect gas

The two equations to be satisfied are.

(*a*) The *substance* equation, i.e. the equation of state for a perfect gas, which is $pV = RT$ or

$$\frac{dp}{p}+\frac{dV}{V} = \frac{dT}{T}. \tag{A. 4.1}$$

(*b*) The process equation, in this case the general polytropic process, defined by

$$\frac{dp}{p}+n\frac{dV}{V} = 0. \tag{A. 4.2}$$

Substitution of (A. 4.2) into (A. 4.1) gives

$$\frac{dp}{p} = \frac{n}{n-1}\frac{dT}{T}, \tag{A. 4.3}$$

and

$$\frac{dV}{V} = -\frac{1}{n-1}\frac{dT}{T}. \tag{A. 4.4}$$

Equations (A. 4.3) and (A. 4.4) are useful in relating pressure and volume respectively direct to temperature, and vice versa.

Now we recall that for a perfect gas

$$dS = \frac{dU+pdV}{T} = c_v\frac{dT}{T}+\frac{pV}{T}\frac{dV}{V} \tag{A. 4.5}$$

$$= c_v\frac{dT}{T}+R\frac{dV}{V}. \tag{A. 4.6}$$

Substitute from (A. 4.4) into (A. 4.6), and we obtain

$$dS = \left(c_v-\frac{R}{n-1}\right)\frac{dT}{T}. \tag{A. 4.7}$$

It is usual to define

$$\gamma = \frac{c_p}{c_v}. \tag{A. 4.8}$$

$$\therefore \quad R = c_p - c_v = c_v(\gamma-1);$$

$$\therefore \quad dS = \frac{c_v(n-\gamma)}{n-1}\frac{dT}{T}.$$

We note from (A. 4.9) that for a perfect gas the isentropic process will be characterised by

$$n = \gamma. \tag{A. 4.10}$$

For the general polytropic process with a perfect gas we have the relations, from (A. 4.3) and (A. 4.4)

$$\left.\begin{aligned}\frac{p_2}{p_1} &= \left(\frac{T_2}{T_1}\right)^{n/(n-1)}, \\ \frac{T_2}{T_1} &= \left(\frac{p_2}{p_1}\right)^{(n-1)/n};\end{aligned}\right\} \tag{A. 4.11}$$

$$\left.\begin{aligned}\frac{V_2}{V_1} &= \left(\frac{T_1}{T_2}\right)^{1/(n-1)}, \\ \frac{T_2}{T_1} &= \left(\frac{V_1}{V_2}\right)^{n-1}.\end{aligned}\right\} \tag{A. 4.12}$$

Further, since

$$dH = c_p\,dT, \tag{A. 4.13}$$

$$H_2 - H_1 = c_p(T_2 - T_1) = c_p T_1\left[\left(\frac{p_2}{p_1}\right)^{\frac{n-1}{n}} - 1\right], \tag{A. 4.14}$$

and

$$S_2 - S_1 = \frac{c_v(n-\gamma)}{n-1}\log\left(\frac{T_2}{T_1}\right). \tag{A. 4.15}$$

Other integrations can obviously be made to express the results in terms of V or p, etc., as required.

We also have

$$S_2 - S_1 = \frac{c_v(n-\gamma)}{n}\log\left(\frac{p_2}{p_1}\right), \tag{A. 4.16}$$

and

$$S_2 - S_1 = c_v(n-\gamma)\log\left(\frac{V_1}{V_2}\right). \tag{A. 4.17}$$

For an expansion, $\log(V_1/V_2)$ is negative. Hence, if the process is adiabatic n must be less than γ, since $S_2 - S_1$ must be positive. Thus for an expansion n can only be greater than or equal to γ if energy is extracted by heating (cooling) during the process, i.e. if Q_t is sufficiently negative.

In a compression, $\log(p_2/p_1)$ is positive, so that n must be greater than γ in an adiabatic process to give increases of entropy. It could only be equal to or less than γ if again we have cooling, i.e. Q_t sufficiently negative.

Finally it is worth noting that in an isothermal process, equation (A. 4.15) becomes inoperable since $T_2/T_1 = 1$ and $n = 1$. We use then either (A. 4.16) or (A. 4.17), which give

$$S_2 - S_1 = -c_v(\gamma - 1)\log\frac{p_2}{p_1} = -R\log\left(\frac{p_2}{p_1}\right),$$

or

$$S_2 - S_1 = R\log\left(\frac{V_2}{V_1}\right). \qquad \text{(A. 4.18)}$$

Thus isothermal expansion is an increase of entropy, while isothermal compression is a decrease of entropy.

A. 5 Polytropic index and mechanical efficiency

We can now apply these relations to analyse the actual performance of expansion and compression processes. The means are now available to obtain the required quantities derived in Section A. 1.

Recalling equation (A. 1.10)

$$\eta_{ae} = \frac{\Delta Q_{ta} - dH_a}{T_a dS_a - dH_a} = \frac{\Delta W_{da}}{T_a dS_a - dH_a}. \qquad \text{(A. 5.1)}$$

Using equations (A. 4.9) and (A. 4.13) gives

$$\eta_{ae} = \frac{\Delta W_{da}}{\left[\frac{c_v(n-\gamma)}{n-1} - c_p\right] dT_a}. \qquad \text{(A. 5.2)}$$

$$\therefore \quad \eta_{ae} = -\frac{\Delta W_{da}(n-1)}{c_v n(\gamma-1)\, dT_a}; \qquad \text{(A. 5.3)}$$

$$\therefore \quad \Delta W_{da} = -\frac{n(\gamma-1)}{\gamma(n-1)}\eta_{ae}\, c_p\, dT_a; \qquad \text{(A. 5.4)}$$

$$\therefore \quad \frac{\Delta W_{da}}{dH_a} = -\frac{n(\gamma-1)}{\gamma(n-1)}\eta_{ae}. \qquad \text{(A. 5.5)}$$

For an adiabatic process, $\Delta W_{da} = -dH_a$, hence

$$\frac{n(\gamma-1)}{\gamma(n-1)}\eta_{ae} = 1. \qquad \text{(A. 5.6)}$$

Thus in an adiabatic process

$$\eta_{ae} = \bar{\eta}_{ae} = \frac{\gamma(n-1)}{n(\gamma-1)}. \qquad \text{(A. 5.7)}$$

It is useful to see this in the form

$$\frac{n-1}{n} = \eta_{ae}\left(\frac{\gamma-1}{\gamma}\right). \qquad \text{(A. 5.8)}$$

$(n-1)/n$ is the exponent of the pressure ratio to give the temperature change. The factor η being less than unity means that the temperature ratio is *not* so much reduced for a given pressure reduction as it would have been in an isentropic adiabatic expansion.

For any flow expansion process with a perfect gas, we can derive the mechanical efficiency and frictional losses from the relation

$$\eta_{ae} = \frac{-\gamma(n-1)}{n(\gamma-1)} \frac{\Delta W_{da}}{dH_a}. \tag{A. 5.9}$$

Turning now to the non-flow expansion process,

$$\eta_{ae} = \frac{\Delta Q_{ta} - dU_a}{T_a dS_a - dU_a} = \frac{\Delta W_{ta}}{\left[\frac{c_v(n-\gamma)}{n-1} - c_v\right] dT_a};$$

$$\therefore \quad \eta_{ae} = -\frac{\Delta W_{ta}}{c_v dT_a}\left(\frac{n-1}{\gamma-1}\right)$$

$$= -\frac{\Delta W_{ta}}{dU_a}\left(\frac{n-1}{\gamma-1}\right); \tag{A. 5.10}$$

$$\therefore \quad \frac{\Delta W_{ta}}{dU_a} = -\eta_{ae}\left(\frac{\gamma-1}{n-1}\right). \tag{A. 5.11}$$

For an adiabatic process $\Delta W_{ta} = -dU$, and

$$\eta_{ae}\left(\frac{\gamma-1}{n-1}\right) = 1, \tag{A. 5.12}$$

or

$$\eta_{ae} = \frac{n-1}{\gamma-1}. \tag{A. 5.13}$$

For any non-flow expansion process with a perfect gas

$$\eta_{ae} = -\left(\frac{n-1}{\gamma-1}\right)\frac{\Delta W_{ta}}{dU_a}. \tag{A. 5.14}$$

The reader may care to satisfy himself that for compression we have the following relations

Flow process

$$\frac{dH_a}{\Delta W_{da}} = -\frac{\gamma(n-1)}{n(\gamma-1)}\eta_{ac}, \quad \text{(A. 5.15) (cf. (A. 5.5))}$$

$$\eta_{ac} = -\frac{n(\gamma-1)}{\gamma(n-1)}\frac{dH_a}{\Delta W_{da}}; \quad \text{(A. 5.16) (cf. (A. 5.9))}$$

Non-flow process

$$\frac{dU_a}{\Delta W_{ta}} = -\left(\frac{n-1}{\gamma-1}\right)\eta_{ac}, \quad \text{(A. 5.17) (cf. (A. 5.11))}$$

$$\eta_{ac} = -\left(\frac{\gamma-1}{n-1}\right)\frac{dU_a}{\Delta W_{ta}}. \quad \text{(A. 5.18) (cf. (A. 5.14))}$$

And that for adiabatic compression we have for a flow process

$$\eta_{ac} = \frac{n(\gamma - 1)}{\gamma(n-1)}, \qquad \text{(A. 5.19) (cf. (A. 5.10))}$$

and for a non-flow process $\eta_{ac} = \dfrac{\gamma - 1}{n - 1}$. (A. 5.20) (cf. (A. 5.19))

To help in the use and recollection of these formulae the following pointers should be borne in mind.

(1) An expansion process produces *output* transmission of work from a *fall* in enthalpy or internal energy. Thus ΔW_d or ΔW_t is positive, and dH or dU is negative.

(2) A compression process produces a *rise* in enthalpy or internal energy from an *input* transmission of work. Hence dH or dU is positive and ΔW_d or ΔW_t is negative.

(3) Thus in *all* cases, ratios $\Delta W_d/dH$ or $\Delta W_t/dU$ or their reciprocals must be *negative*.

(4) We may think of the ratios $\Delta W_d/dH$ or $\Delta W_t/dU$ for expansion processes or their reciprocals for compression processes, as representing a kind of *overall* ratio formed of the mechanical efficiency η_a multiplied by a factor which is a function of n and γ and which allows for the effect of heating or cooling.

(5) Thus the only equations which we need attempt to keep in our minds are (A. 5.5), (A. 5.11), (A. 5.15) and (A. 5.17) which are repeated here:

Expansion

$$\text{flow} \quad \frac{\Delta W_{da}}{dH_a} = -\frac{n(\gamma-1)}{\gamma(n-1)}\eta_{ae}; \qquad \text{(A. 5.5)}$$

$$\text{non-flow} \quad \frac{\Delta W_{ta}}{dU_a} = -\left(\frac{\gamma-1}{n-1}\right)\eta_{ae}. \qquad \text{(A. 5.11)}$$

Compression

$$\text{flow} \quad \frac{dH_a}{\Delta W_{da}} = -\frac{\gamma(n-1)}{n(\gamma-1)}\eta_{ac}; \qquad \text{(A. 5.15)}$$

$$\text{non-flow} \quad \frac{dU_a}{\Delta W_{ta}} = -\left(\frac{n-1}{\gamma-1}\right)\eta_{ac}. \qquad \text{(A. 5.17)}$$

All the others can easily be obtained from these. It must be recalled that $\Delta Q_t = dH + \Delta W_d$ for flow process or $\Delta Q_t = dU + \Delta W_t$ for non-flow processes, with complete generality, so that the left-hand side of each equation is negative unity if there is no heating nor cooling. Thus we easily obtain the adiabatic formulae. If ΔQ_t is not zero, the left-hand side of each equation is expressible in the same way as we now illustrate for equation (A. 5.5).

$$\frac{\Delta W_{da}}{dH_a} = \frac{\Delta W_{da}}{\Delta Q_{ta} - \Delta W_{da}} = \frac{1}{\dfrac{\Delta W_{da}}{\Delta Q_{ta}} - 1}$$

$$= \frac{\Delta Q_{ta} - dH_a}{dH_a} = \frac{\Delta Q_{ta}}{dH_a} - 1. \qquad \text{(A. 5.21)}$$

Equations (A. 5.21) show how we can relate the R.H.S. of each equation either to the ratio of transmitted work to enthalpy change, of transmitted work to transmitted heat, or of transmitted heat to enthalpy change.

As a further mnemonic it may be noted that the compression equation can be written *exactly* as the expansion equations if we use $1/\eta_{ac}$ thus:

Compression

$$\text{flow} \qquad \frac{\Delta W_{da}}{dH_a} = -\left[\frac{n(\gamma-1)}{\gamma(n-1)}\right]\frac{1}{\eta_{ac}}; \qquad \text{(A. 5.22)}$$

$$\text{non-flow} \qquad \frac{\Delta W_{ta}}{dU_a} = -\left(\frac{\gamma-1}{n-1}\right)\frac{1}{\eta_{ac}}. \qquad \text{(A. 5.23)}$$

Thus the form is identical with (A. 5.5) and (A. 5.11) except for the reciprocal efficiency.

A. 6 Adiabatic processes. Further discussion of path and isentropic efficiencies

Consider an adiabatic flow expansion process from p_1, V_1 as illustrated in Figure A. 6.1.

The expansion is from p_1 to p_2 which are fixed by external controls, and the actual thermodynamic path is along aa' from V_1 to V_2. If there were no friction, the ideal path (isentropic since adiabatic and frictionless) would be from V_1 to V_s.

The amount of the frictional losses is indicated by the area $aa'l$ so that the actual work done is indicated by the area $balc$ although as noted previously (p. 133) the line al has no real significance on a state diagram.

We have seen that for a perfect gas the enthalpy change for a given temperature change is

$$dH = c_p\, dT. \qquad \text{(A. 6.1)}$$

In an adiabatic expansion, $\Delta W_d = -dH$,

$$\therefore \quad \Delta W_d = -c_p\, dT. \qquad \text{(A. 6.2)}$$

For the actual path,

$$\Delta W_{da} = -c_p\, dT_a. \qquad \text{(A. 6.3)}$$

The work potential of an element of the actual path is $-V_a\,dp_a$ which is $-dH_a+T_a\,dS_a$. Using (A. 6.1) and (A. 4.9) gives

$$-V_a\,dp_a = -c_p\,dT_a + \frac{c_v(n-\gamma)}{n-1}\,dT_a$$

$$= c_p\,\frac{n(\gamma-1)}{\gamma(n-1)}\,dT_a. \qquad \text{(A. 6.4)}$$

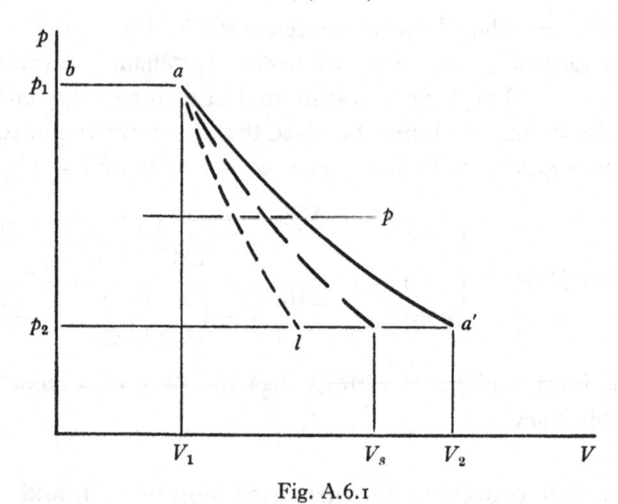

Fig. A.6.1

Dividing (A. 6.3) by (A. 6.4) gives the path efficiency η_{ae} as before:

$$\eta_{ae} = \frac{\gamma(n-1)}{n(\gamma-1)}. \qquad \text{(A. 6.5)}$$

Now for the ideal isentropic path, the ideal element of work may be indicated by

$$\Delta W_s = -V_s\,dp_s = -c_p\,dT_s \qquad \text{(A. 6.6)}$$

where the suffix s indicates the condition of constant entropy.

Hence the isentropic efficiency η_{se} is

$$\eta_{se} = \frac{\Delta W_{da}}{\Delta W_s} = dT_a/dT_s. \qquad \text{(A. 6.7)}$$

The solidus line is used for dT_a/dT_s to remind us that it is a ratio of elements of temperature drop, and not a differential of one temperature with respect to the other.

But since aa' is a polytropic with index n and as is a polytropic with index γ (constant entropy), we have, from equation (A. 4.3),

$$\left.\begin{aligned} dT_a &= \left(\frac{n-1}{n}\right) T_a \frac{dp}{p}; \\ dT_s &= \left(\frac{\gamma-1}{\gamma}\right) T_s \frac{dp}{p}. \end{aligned}\right\} \tag{A. 6.8}$$

Thus at any pressure p during the path (see line p–p on Figure A. 6.1) the isentropic efficiency is

$$\eta_{se} = \frac{\gamma(n-1)}{n(\gamma-1)} \frac{T_a}{T_s}. \tag{A. 6.9}$$

Now we notice a very interesting point. The path efficiency of an elementary change dp as shown by equation (A. 6.5), is *constant* for every pressure point in any polytropic process, so that as assumed in the discussions of Section A. 5, it can be used directly for a whole finite compression or expansion change such as that from p_1 to p_2. But the isentropic efficiency of an elementary change dp is not constant. By applying equations (A. 4.11) and since each path starts from the same temperature it can be shown that

$$\frac{T_a}{T_s} = \left(\frac{p_1}{p}\right)^{(\gamma-n)/n\gamma},$$

so that

$$\eta_{se} = \frac{\gamma(n-1)}{n(\gamma-1)} \left(\frac{p_1}{p}\right)^{(\gamma-n)/n\gamma} \tag{A. 6.10}$$

$$= \eta_{ae} \left(\frac{p_1}{p}\right)^{(\gamma-n)/n\gamma}. \tag{A. 6.11}$$

It was shown in Section A. 5 that for an adiabatic expansion n must be $< \gamma$, hence the index of p_1/p is positive and so η_{se} for an element starts immediately equal to η_{ae} but then gradually increases as the expansion takes place. This is another obvious disadvantage of the concept of isentropic efficiency.

In practice, of course, what one uses is the mean efficiency throughout the finite expansion from p_1 to p_2. For path efficiency we have simply, since it is constant,

$$\bar{\eta}_{ae} = \eta_{ae} = \frac{\gamma(n-1)}{n(\gamma-1)}. \tag{A. 6.12}$$

For isentropic efficiency we have to take the finite temperature change ratio $(T_2 - T_1)/(T_{2s} - T_1)$ instead of equation (A. 6.7). We have

$$T_2 - T_1 = T_1 \left[\left(\frac{p_1}{p}\right)^{(n-1)/n} - 1\right], \tag{A. 4.11}$$

and
$$\bar{\eta}_{se} = \frac{1-\left(\dfrac{p}{p_1}\right)^{(n-1)/n}}{1-\left(\dfrac{p}{p_1}\right)^{(\gamma-1)/\gamma}}. \qquad \text{(A. 6.13)}$$

The reader should try the exercise of determining the corresponding expressions for non-flow expansion, and for flow and non-flow compression.

Finally a word should be said about the practical utility of $\bar{\eta}_a$ and $\bar{\eta}_s$. It should be remembered that the whole objective of all engineering theoretical work is to assist in design, and to assist in the understanding and interpretation of operational data. Almost certainly no actual machine, running adiabatically, operates such that the polytropic exponent n is actually constant. Equally certainly no actual elemental 'efficiency' is constant, nor is it likely to change in an absolutely regular fashion. The conceptual structure we set up is to enable us to handle data, to interpret observations, and all our practice should regard η_a, η_s, $\bar{\eta}_a$, $\bar{\eta}_s$, n, etc., simply as tools with which to do such analysis and interpretation. Hence we use whichever form and approach is most convenient for the particular purpose in hand. Sometimes path efficiency is most convenient, sometimes isentropic efficiency. Wherever the perfect gas assumptions are reasonably tenable, path efficiency procedure is usually preferable.

A. 7 The temperature/entropy diagram

Throughout the main body of the book we have discussed processes in terms of the pressure/volume diagram. But many useful techniques are based on the use of the temperature/entropy diagram, the origin and nature of which can be described as follows.

The basic equation is

$$\begin{aligned}\Delta Q_t &= dU + p\,dV - \Delta W_f \\ &= T\,dS - \Delta W_f. \end{aligned} \qquad \text{(A. 7.1)}$$

Now since U is a property, $\oint dU = 0$, and hence

$$\oint T\,dS = \oint p\,dV; \qquad \text{(A. 7.2)}$$

$$\oint \Delta \bar{Q}_t = \oint T\,dS - \oint \Delta W_f = \oint \Delta W_t. \qquad \text{(A. 7.3)}$$

Thus a closed cycle on a diagram with T and S as ordinate and abscissa respectively is equal in its energy area to the corresponding

p, V diagram. Figure A. 7.1 shows the corresponding diagrams for the case discussed in Chapter 3, Section 3.5, Figure 3.5.1.

The case represented is for a real cycle, including friction. Hence for the portions in which $\Delta Q_t = 0$, $T dS = \Delta W_f$ and must be positive. Thus the T, S diagram shows the entropy increase clearly in each of these steps. Where energy is being supplied or rejected by heat at the constant temperatures T_s and T_r we have

$$Q_s = T_s(S_2 - S_1) - W_{f_{12}}; \tag{A. 7.4}$$

$$Q_r = T_r(S_3 - S_4) + W_{f_{34}}. \tag{A. 7.5}$$

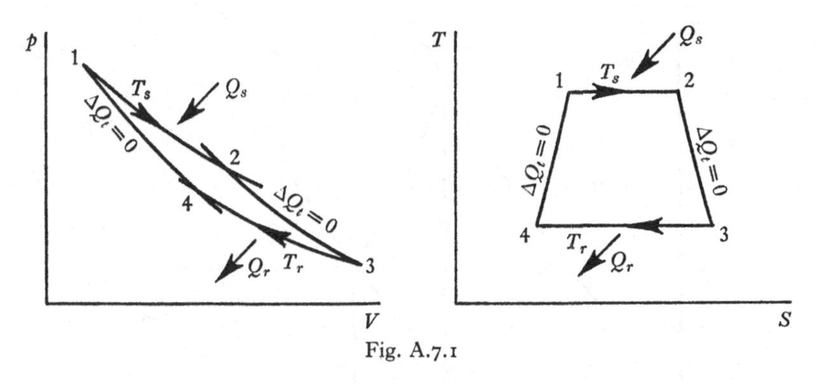

Fig. A.7.1

The diagram shows graphically that $S_3 - S_4$ is necessarily greater than $S_2 - S_1$.

Similarly the discussion of any part of a process on a single substance may usefully be facilitated by the T, S diagram. Consider for example an adiabatic expansion process from p_1 to p_2 such as shown in Figure A. 2.1 of this Appendix. The corresponding T, S diagram is that of Figure A. 7.2.

The interesting point is that on this diagram we do in fact find an area equal to the frictional work. Since it is an adiabatic process, $\Delta Q_t = 0$,

$$\left.\begin{aligned} T dS - \Delta W_f &= 0, \\ \int_1^2 T dS &= \Delta W_{f_{12}}. \end{aligned}\right\} \tag{A. 7.6}$$

On the diagram AA' is the actual process path. A vertical line from the initial condition is an isentropic. Thus the area $aAA'a'$ is equal to the energy dissipation $W_{f_{12}}$.

A further point of note is that where we are considering a process in which kinetic and potential energies are negligible we have a complete correspondence between total work, pressure, and volume on the one hand, and heat, temperature, and entropy on the other. For we have

$$\Delta W_t = p\,dV - \Delta W_f, \tag{A. 7.7}$$

and

$$\Delta Q_t = T\,dS - \Delta W_f. \tag{A. 7.8}$$

These give of course

$$\Delta Q_t - \Delta W_t = T\,dS - p\,dV = dU. \tag{A. 7.9}$$

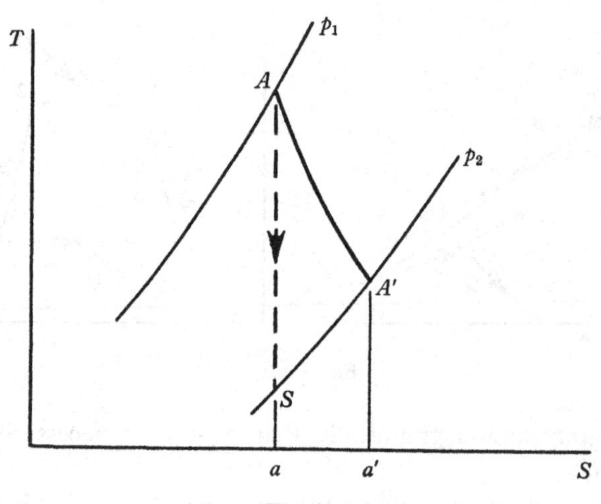

Fig. A.7.2

It is however worthwhile emphasising that equation (A. 7.8) is true even when kinetic and potential energies are appreciable. These terms only appear in the work equation (A. 7.7), and of course remain in (A. 7.9).

A. 8 The enthalpy/entropy diagram

Another diagram in frequent use is that in which the enthalpy H is plotted against entropy S. Its utility arises from the fact that, from equation (A. 1.3),

$$\Delta W_a = -V\,dp + \Delta Q_t - T\,dS - d(\tfrac{1}{2}v^2) - d\Phi.$$

Where potential energy is negligible this may be written

$$\Delta W_a + d(\tfrac{1}{2}v^2) = \Delta Q_t - dH. \tag{A. 8.1}$$

THE ENTHALPY/ENTROPY DIAGRAM

The H, S diagram is of particular use for adiabatic processes, since we then have simply

$$\Delta W_d + d(\tfrac{1}{2}v^2) = -dH. \qquad \text{(A. 8.2)}$$

Thus if we have the H, S diagram for a working substance as in Figure A. 8.1, a vertical line ab on it will represent the sum of dynamic work done and kinetic energy gained in an isentropic expansion from p_1 to p_2 as shown. This could occur adiabatically only if the process were ideally frictionless. An actual adiabatic process will show the line aa'. The actual enthalpy drop is thus $H_a - H'_a$ instead of the ideal $H_a - H_b$.

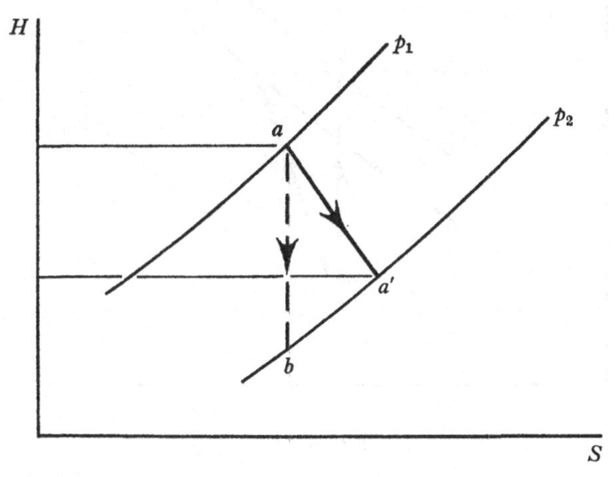

Fig. A.8.1

When the fluid flows through a stationary nozzle, ΔW_d is zero and the whole of $H_a - H'_a$ goes to the creation of kinetic energy. Conversely Figure A. 8.2 illustrates the conversion of kinetic energy to enthalpy in a stationary diffuser.

The same conversion $H_b - H_a$ succeeds only in gaining a compression from p_1 to p'_2 instead of to p_2.

The H, S diagram and T, S diagrams are conveniently used for substances which are not perfect gases and for which the techniques based on the perfect gas – as discussed in Sections A. 1 to A. 7 of this Appendix – are not sufficiently good approximations. But of course they may be used also to illustrate graphically processes operating on any substance. Since for a perfect gas $dH/dT = C_p$ and C_p is constant, the H, S and T, S diagrams are identical for a perfect gas, except for a dimensional scale constant on the ordinate.

When both ΔW_d and $d(\frac{1}{2}v^2)$ are zero, equation (A. 8. 2) shows that then

dH is zero also. This is what is called a throttling process. One example is when a fluid flows through an insulated porous plug so that neither work is done nor velocity gained. We then have

$$\Delta Q_t = 0, \quad T\,dS = \Delta W_f;$$

$$dH = 0, \quad T\,dS = -V\,dp.$$

The pressure fall is used entirely to overcome friction.

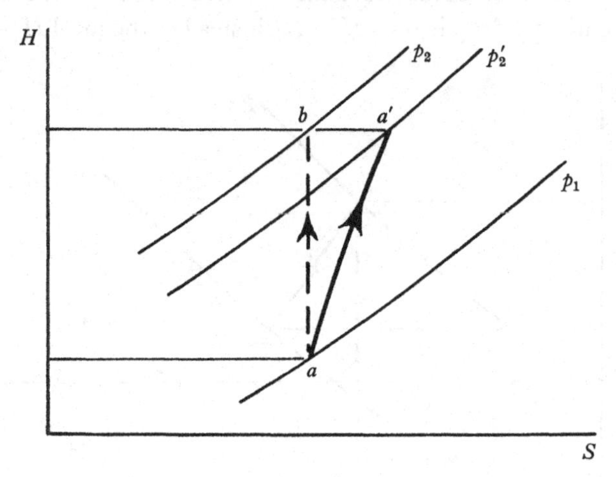

Fig. A.8.2

An equivalent case occurs integrally when a fluid flows through an orifice from a wide area pipe into another wide area pipe, so that its initial and final velocities are negligible, and no work is done. If the whole apparatus is also insulated the process is again adiabatic. We now have therefore

$$-\int_1^2 dH = \int_1^2 \Delta W_d + \int_1^2 d(\tfrac{1}{2}v^2) = 0,$$

i.e. $$H_2 = H_1, \quad \text{and} \quad \int T\,dS = -\int V\,dp.$$

In the particular case of a perfect gas H is dependent on temperature only and independent of pressure and so the temperature after throttling is the same as before. Hence measurement of temperature change, if any, in a throttling experiment can be used to see whether or not a substance behaves as a perfect gas.

SUBJECT INDEX